Graph Theory

Mathematical Olympiad Series

ISSN: 1793-8570

Series Editors: Lee Peng Yee *(Nanyang Technological University, Singapore)*
Xiong Bin *(East China Normal University, China)*

Published

Vol. 1   A First Step to Mathematical Olympiad Problems
by Derek Holton (University of Otago, New Zealand)

Vol. 2   Problems of Number Theory in Mathematical Competitions
by Yu Hong-Bing (Suzhou University, China)
translated by Lin Lei (East China Normal University, China)

Vol. 3   Graph Theory
by Xiong Bin (East China Normal University, China) &
  Zheng Zhongyi (High School Attached to Fudan University, China)
translated by Liu Ruifang, Zhai Mingqing & Lin Yuanqing
  (East China Normal University, China)

Xiong Bin
East China Normal University, China

Zheng Zhongyi
High School Attached to Fudan University, China

Vol. 3 | Mathematical Olympiad Series

Graph Theory

translated by

Liu Ruifang
Zhai Mingqing
Lin Yuanqing
East China Normal University, China

Published by

East China Normal University Press
3663 North Zhongshan Road
Shanghai 200062
China

and

World Scientific Publishing Co. Pte. Ltd.
5 Toh Tuck Link, Singapore 596224
USA office: 27 Warren Street, Suite 401-402, Hackensack, NJ 07601
UK office: 57 Shelton Street, Covent Garden, London WC2H 9HE

British Library Cataloguing-in-Publication Data
A catalogue record for this book is available from the British Library.

GRAPH THEORY
Mathematical Olympiad Series — Vol. 3

Copyright © 2010 by East China Normal University Press and
World Scientific Publishing Co. Pte. Ltd.

All rights reserved. This book, or parts thereof, may not be reproduced in any form or by any means, electronic or mechanical, including photocopying, recording or any information storage and retrieval system now known or to be invented, without written permission from the Publisher.

For photocopying of material in this volume, please pay a copying fee through the Copyright Clearance Center, Inc., 222 Rosewood Drive, Danvers, MA 01923, USA. In this case permission to photocopy is not required from the publisher.

ISBN-13 978-981-4271-12-7 (pbk)
ISBN-10 981-4271-12-8 (pbk)

Printed in Singapore by B & Jo Enterprise Pte Ltd

Editors
XIONG Bin *East China Normal University, China*
Lee Peng Yee *Nanyang Technological University, Singapore*

Original Authors
XIONG Bin *East China Normal University, China*
ZHENG Zhongyi *High School Affiliated to Fudan University, China*

English Translators
LIU Ruifang *East China Normal University, China*
ZHAI Mingqing *East China Normal University, China*
LIN Yuanqing *East China Normal University, China*

Copy Editors
NI Ming *East China Normal University Press, China*
ZHANG Ji *World Scientific Publishing Co., Singapore*
WONG Fook Sung *Temasek Polytechnic, Singapore*
KONG Lingzhi *East China Normal University Press, China*

Introduction

In 1736, Euler founded Graph Theory by solving the Königsberg seven-bridge problem. It has been more than two hundred years till now. Graph Theory is the core content of Discrete Mathematics, and Discrete Mathematics is the theoretical basis of Computer Science and Network Information Science. This book vulgarly introduces in an elementary way some basic knowledge and the primary methods in Graph Theory. Through some interesting mathematic problems and games the authors expand the knowledge of Middle School Students and improve their skills in analyzing problems and solving problems.

Contents

Introduction		Vii
Chapter 1	Definition of Graph	1
Chapter 2	Degree of a Vertex	13
Chapter 3	Turán's Theorem	24
Chapter 4	Tree	40
Chapter 5	Euler's Problem	51
Chapter 6	Hamilton's Problem	63
Chapter 7	Planar Graph	75
Chapter 8	Ramsey's Problem	84
Chapter 9	Tournament	101
Solutions		110
Index		145

Chapter 1   Definition of Graph

Graph theory is a branch of mathematics on the study of graphs. The graph we consider here consists of a set of points together with lines joining certain pairs of these points. The graph represents a set that has binary relationship.

In recent years, graph theory has experienced an explosive growth and has generated extensive applications in many fields.

We often encounter the following phenomena or problems:

In a group of people, some of them know each other, but others do not.

There are some cities. Some pairs of them are connected by airlines and others are not.

There is a set of points in the plane. The distance between some of them is one and others are not one.

All the above phenomena or problems contain two aspects: one is object, such as people, football teams, cities, points and so on; and the other is a certain relationship between these objects, such as "knowing each other", "having a contest", "the distance between" and so on. In order to represent these objects and the relationships, we could use a point as an object, which is called a *vertex*. If any two objects have a relationship, then there is a line joining them, which is called an *edge*. Then we have constructed a graph.

We call the figure a *graph*[a]. For instance, the three graphs G_1,

---

a) The general definition of graphs: a graph is a triplet (V, E, ψ), where V and E are two disjoint sets, V is nonempty and ψ is a mapping from $V \times V$ to E. The sets V, E, ψ are vertex set, edge set and incidence function, respectively.

G_2, G_3 in Fig. 1.1 are isomorphic, which contain some vertices and edges joining them, representing some objects and the relationships between them.

Fig. 1.1 shows three graphs G_1, G_2, G_3, where vertices are represented by small circles.

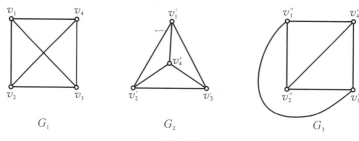

Fig. 1.1

We can see that in the definition of graphs there are no requirement on the location of the vertices, the length and the curvature of the edges, and the fact whether the vertices and the edges are in the same plane or not. However, we do not allow an edge passing through the third vertex and also not let an edge intersect itself. In graph theory, if there is a bijection from the vertices of G to the vertices of G' such that the number of edges joining v_i and v_j equals the number of edges joining v_i' and v_j', then two graphs G and G' are *isomorphic* and considered as the same graph.

A graph $G' = (V', E')$ is called a *subgraph* of a graph $G = (V, E)$ if $V' \subseteq V$, $E' \subseteq E$, that is, all the vertices of G' are the vertices of G and the edges of G' are the edges of G.

For instance, the graphs G_1, G_2 in Fig. 1.2 are the subgraphs of G.

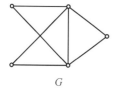 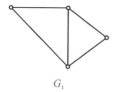 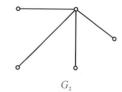

Fig. 1.2

If there is an edge joining v_i and v_j in graph G, then v_i and v_j are *adjacent*. Otherwise, they are nonadjacent. If the vertex v is an end of the edge e, then v is *incident* to e. In Fig. 1.3, v_1 and v_2 are adjacent, but v_2 and v_4 are not. The vertex v_3 is incident to the edge e_4.

We called the edge a *loop* if there is an edge joining the vertex and itself. For instance, the edge e_6 in Fig. 1.3 is a loop.

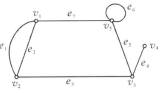

Fig. 1.3

Two or more edges with the same pair of ends are called *parallel edges*. For instance, the edges e_1, e_2 in Fig. 1.3 are parallel edges.

A graph is *simple* if it has no loops or parallel edges. The graphs G_1, G_2, G_3 in Fig. 1.1 are simple, whereas the graph in Fig. 1.3 is not. In a simple graph, the edge joining v_i and v_j is denoted by (v_i, v_j). Certainly, (v_i, v_j) and (v_j, v_i) are considered as the same edge.

A *complete* graph is a simple graph in which any two vertices are adjacent. We denote the complete graph with n vertices by K_n. The graphs K_3, K_4, K_5 in Fig. 1.4 are all complete graphs. The number of edges of the complete graph K_n is $\binom{n}{2} = \frac{1}{2}n(n-1)$.

K_3 K_4 K_5

Fig. 1.4

A graph is *finite* if both the number of the vertices $|V|$ ($|V|$ is also said to be the order of G) and the number of edges $|E|$ are finite. A graph is *infinite* if $|V|$ or $|E|$ is infinite.

In this chapter, unless specified, all graphs under discussion should be taken to be finite simple graphs.

These fundamental concepts mentioned above help us to consider and solve some questions.

Example 1 There are 605 people in a party. Suppose that each of them shakes hands with at least one person. Prove that there must be someone who shakes hands with at least two persons.

Proof We denote the 605 people by 605 vertices v_1, v_2, ..., v_{605}. If any two of them shake hands, then there is an edge joining the corresponding vertices.

In this example we are going to prove that there must be someone who shakes hands with at least two persons. Otherwise, each of them shakes hands with at most one person. Moreover, according to the hypothesis each of them shakes hands with at least one person. Thus we have each of them just shakes hands with one person. It implies that the graph G consists of several figures that every two vertices are joined by only one edge.

Suppose that G have r edges. So G has $2r$ (even) vertices. It contradicts the fact that the number of vertices of G is 605 (odd).

We complete the proof.

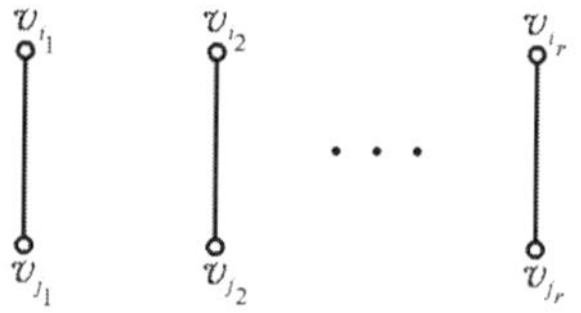

Fig. 1.5

Example 2 Is it possible to change the state in Fig. 1.6 to the state in Fig. 1.7 by moving the knights several times? (In the figures, W stands for white knight, and B stands for black knight. knight should be moved by following the international chess regulation)

Solution As Fig. 1.8 shows, the nine squares are numbered and each of them is represented by a vertex in the plane. If the knight can be moved from one square to anther square, then there is an edge joining the two corresponding vertices, as Fig. 1.9 shows.

W		W
B		B

Fig. 1.6

W		B
B		W

Fig. 1.7

1	4	7
2	5	8
3	6	9

Fig. 1.8

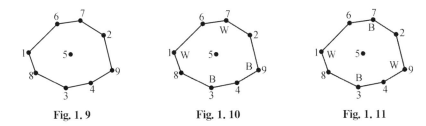

Fig. 1.9   Fig. 1.10   Fig. 1.11

Thus the beginning state in Fig. 1.6 and the state in Fig. 1.7 are represented by the two graphs as in Fig. 1.10, Fig. 1.11, respectively.

Obviously, the order of the knight on the circle cannot be changed from the state that two white knight are followed by two white knight into the state that white knight and black knight are interlaced. So it is impossible to change the states as required.

Example 3 There are n people $A_1, A_2, \ldots, A_n$ taking part in a mathematics contest, where some people know each other and any two people who do not know each other would have common acquaintance. Suppose that A_1 and A_2 know each other, but do not have common acquaintance. Prove that the acquaintances of A_1 are as many as those of A_2.

Proof Denote the n people $A_1, A_2, \ldots, A_n$ by n vertices $v_1, v_2, \ldots, v_n$. If two people know each other, then there is an edge joining the two corresponding vertices. Then we get a simple graph G. The vertices of G satisfy that any two nonadjacent vertices have a common neighbor. We shall prove two adjacent vertices v_1 and v_2 have the same number of neighbors.

The set of neighbors of the vertex v_1 is denoted by $N(v_1)$ and the set of neighbors of the vertex v_2 is denoted by $N(v_2)$. If there is a vertex v_i in $N(v_1)$ and $v_i \neq v_2$, then v_i is not in $N(v_2)$. Otherwise A_1 and A_2 have the common acquaintance A_i. Thus v_2 and v_i have a common neighbor v_j and $v_j \neq v_1$. So $N(v_2)$ contains v_j, as Fig. 1.12 shows. For v_i, v_k in $N(v_1)$, which are

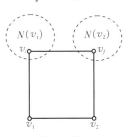

Fig. 1.12

distinct from v_2, both of them cannot be adjacent to a vertex v_j in $N(v_2)$, which is distinct from v_1. Otherwise, two nonadjacent vertices v_1, v_j have three common neighbors v_2, v_i, v_k. Therefore v_k in $N(v_1)$, which is distinct from v_k, must have a neighbor v_l in $N(v_2)$, which is distinct from v_j. So the number of vertices in $N(v_1)$ is not greater than that of $N(v_2)$. Similarly the number of vertices in $N(v_2)$ is not greater than that of $N(v_1)$. Thus the edges incident to v_1 are as many as those incident with v_2.

Example 4 Nine mathematicians meet at an international mathematics conference. For any three persons, at least two of them can have a talk in the same language. If each mathematician can speak at most three languages, prove that at least three mathematicians can have a talk in the same language. (USAMO 1978)

Proof Denote the 9 mathematicians by 9 vertices v_1, v_2, ..., v_9. If two of them can have a talk in the ith language, then there is an edge joining the corresponding vertices and color them with the ith color. Then we get a simple graph with 9 vertices and edges colored. Every three vertices have at least one edge joining them and the edges incident to a vertex are colored in at most three different colors. Prove that there are three vertices in graph G, any two of which are adjacent to the three edges colored with the same color. (This triangle is called *monochromatic* triangle.)

If the edges (v_i, v_j), (v_i, v_k) have the ith color, then the vertices v_j, v_k are adjacent and edge (v_j, v_k) has the ith color. Thus for vertex v_1, there are two cases:

(1) The vertex v_1 is adjacent to v_2, ..., v_9. By the pigeonhole principle, at least two edges, without loss of generality, denoted by (v_1, v_2), (v_1, v_3), have the same color. Thus triangle $\triangle v_1 v_2 v_3$ is a monochromatic triangle.

(2) The vertex v_1 is nonadjacent to at least one of v_2, ..., v_9. Without loss of generality, we suppose that v_1 is nonadjacent to v_2. For every three vertices there is at least one edge joining them, so there are at least seven edges from vertices v_3, v_4, ..., v_9 to the

vertex v_1 or v_2. From that we know at least four vertices of v_3, v_4, ..., v_9 are adjacent with vertex v_1 or v_2. Without loss of generality, we suppose that v_3, v_4, v_5, v_6 are adjacent to v_1, as it is shown in Fig. 1.13. Thus there must be two edges of (v_1, v_3), (v_1, v_4), (v_1, v_5), (v_1, v_6) which have the same color. Suppose (v_1, v_3), (v_1, v_4) have the same color, then $\triangle v_1 v_3 v_4$ is a monochromatic triangle.

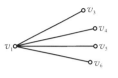

Fig. 1.13

Remark If the number 9 in the question is replaced by 8, then the proposition is not true. Fig. 1.14 gives a counterexample. Denote the 8 vertices by v_1, v_2, ..., v_8 and 12 colors by 1, 2, ..., 12, and there is no monochromatic triangle in the graph.

Fig. 1.14

The following example is the third question of national senior middle school mathematics contest in 2000.

Example 5 There are n people, any two of whom have a talk by telephone at most once. Any $n-2$ of them have a talk by telephone 3^m times, where m is a natural number. Determine the value of n. (China Mathematical Competition)

Solution Obviously $n \geqslant 5$. Denote the n persons by the vertices A_1, A_2, ..., A_n. If A_i, A_j have a talk by telephone, then there is an edge (A_i, A_j). Thus there is an edge joining two of the n vertices. Without loss of generality, we suppose that it is (A_1, A_2).

Suppose there is no edge joining A_1 and A_3. Consider $n-2$ vertices A_1, A_4, A_5, ..., A_n; A_2, A_4, A_5, ..., A_n and A_3, A_4,

$A_5, \ldots, A_n$. We know the number of edges joining any of A_1, A_2, A_3 to all of $A_4, A_5, \ldots, A_n$ is equal and we denote it by k.

Add A_2 to the set $A_1, A_4, A_5, \ldots, A_n$, then there are $S = 3^m + k + 1$ edges joining the $n-1$ vertices. Take away any vertex from $n-1$ vertices, the number of edges joining the remaining $n-2$ vertices is always 3^m. So there are $k+1$ edges joining every vertex and the remaining $n-2$ vertices. Therefore,

$$S = \frac{1}{2}(n-1)(k+1).$$

Similarly, add A_3 to the set $A_1, A_4, A_5, \ldots, A_n$. We get $n-1$ vertices and the number of edges is $t = 3^m + k = \frac{1}{2}(n-1)k$.

For $S = t + 1$, we have

$$\frac{1}{2}(n-1)(k+1) = \frac{1}{2}(n-1)k + 1,$$

that is $n = 3$. A contradiction. Thus there is an edge joining A_1, A_3.

Similarly, there is also an edge joining A_2 and A_3. Moreover, there must be edges joining A_1, A_2 and all A_i ($i = 3, 4, \ldots, n$).

For A_i, A_j ($i \neq j$), there is an edge joining A_i and A_1. So there is an edge joining A_i and A_j. Thus it is a complete graph. Therefore,

$$3^m = \frac{1}{2}(n-2)(n-3).$$

Hence we have $n = 5$.

Example 6 There are n ($n > 3$) persons. Some of them know each other and others do not. At least one of them does not know the others. What is the largest value of the number of persons who know the others?

Solution Construct the graph G: denote the n persons by n vertices and two vertices are adjacent if and only if the two corresponding persons know each other.

For at least one of them does not know the others, in graph G there are at least two vertices which are not adjacent. Suppose that there is no edge $e = (v_1, v_2)$ joining v_1, v_2. Thus G must be $K_n - e$ if it has the most edges. That is the graph taken away an edge e from the complete graph K_n. The largest number of vertices which is adjacent with the remaining vertices is $n - 2$. So the largest number of people who know the others is $n - 2$.

The following example is from the 29th International Mathematical Olympiad (1988).

Example 7 Suppose that n is a positive integer and $A_1, A_2, \ldots, A_{2n+1}$ is a subset of a set B.

Suppose that

(1) each A_i has exactly $2n$ elements;
(2) each $A_i \cap A_j (1 \leq i < j \leq 2n + 1)$ has exactly one element;
(3) each element of B belongs to at least two A_i's.

For which values of n can one assign to every element of B one of the number 0 and 1 in such a way that A_i has 0 assigned to exactly n of its elements?

Solution At first, the words "at least" in (3) can be replaced by "exactly". If there is an element $a_1 \in A_1 \cap A_{2n} \cap A_{2n+1}$, then each of the remaining $2n - 2$ subsets $A_2, A_3, \ldots, A_{2n-1}$ has at most one element of A_1. Thus there is at least one element in A_1 but not in $A_2 \cup A_3 \cup \cdots \cup A_{2n-1} \cup A_{2n} \cup A_{2n+1}$.

It contradicts (3).

Construct the complete graph K_{2n+1}, where every vertex v_i represents a subset A_i and every edge $(v_i, v_j) = e_{ij} (1 \leq i, j \leq 2n+1, i \neq j)$ represents the common element of A_i, A_j. So the question can be changed into: what property does n satisfy such that by as signing the edges of K_{2n+1} to 0 or 1, exactly n edges of the $2n$ edges incident to any vertex v_i are assigned to 0?

K_{2n+1} has $n(2n+1)$ edges. If the required method of assigning can be met, then there are $\frac{1}{2}n(2n+1)$ edges which are assigned to 0. So n must be even.

Conversely, if $n = 2m$ is even, we assign the edges (v_i, v_{i-m}),

$(v_i, v_{i-m+1}), \ldots, (v_i, v_{i-1}), (v_i, v_{i+1}), \ldots, (v_i, v_{i+m}), i = 1, 2, \ldots, 2n + 1$, to 0, otherwise to 1 in K_{2n+1}. Then the method can meet the requirement. (Note that $v_{(2n+1)+i} = v_i$).

Therefore, the condition of the question is satisfied if and only if n is even.

The following problem is from the IMO preseleced questions in 1995.

Example 8 There are $12k$ persons attending a conference. Each of them shakes hands with $3k + 6$ persons, where any two of them shake hands with the same number of people. How many persons are there in the conference?

Solution Suppose that for any two persons, they shake hands with n people. For one person a, the set of all the persons shaking hands with a is denoted by A and the set, the other persons by B. We know from the problem that $|A| = 3k + 6$, $|B| = 9k - 7$. For $b \in A$, n persons shaking hands with a, b are all in A. Therefore, b shakes hands with n persons in A and $3k + 5 - n$ persons in B. For $c \in B$, n persons shaking hands with a, c are all in A. Thus the number of persons in A who have shaken hands with someone in B is

$$(3k+6)(3k+5-n) = (9k-7)n,$$

$$n = \frac{(3k+6)(3k+5)}{12k-1}.$$

So $16n = \dfrac{(12k - 1 + 25)(12k - 1 + 21)}{(12k - 1)}$.

Obviously, $(3, 12k - 1) = 1$. So $(12k - 1) | 25 \times 7$. For $12k - 1$ divided by 4 leaves 3, $12k - 1 = 7, 5 \times 7, 5^2 \times 7$. By calculating $12k - 1 = 5 \times 7$ has the only integer solution $k = 3$, $n = 6$.

Next we construct a figure consists of 36 points. Each point is incident to 15 edges and for any two points there are 6 points adjacent to them.

Naturally, we can use 6 complete graphs K_6. Divide the 36 points into 6 teams and label the points in the same team. We get a 6×6 square matrix

$$\begin{matrix} 1 & 2 & 3 & 4 & 5 & 6 \\ 6 & 1 & 2 & 3 & 4 & 5 \\ 5 & 6 & 1 & 2 & 3 & 4 \\ 4 & 5 & 6 & 1 & 2 & 3 \\ 3 & 4 & 5 & 6 & 1 & 2 \\ 2 & 3 & 4 & 5 & 6 & 1 \end{matrix}$$

For any point in the square matrix, it only connects with 15 points in the same row, in the same column, or having the same label. It is obvious that for any two persons there are 6 persons who have shaken hands with them.

Exercise 1

1 Consider the graph $G = (V, E)$, where $V = \{v_1, v_2, \ldots, v_5\}$, and $E = \{(v_1, v_2), (v_2, v_4), (v_3, v_4), (v_4, v_5), (v_1, v_3)\}$. Draw the graph G.

2 Let G be a simple graph, where $|V| = n$, $|E| = e$. Prove that $e \leqslant \dfrac{n(n-1)}{2}$.

3 Show the following two graphs are isomorphic.

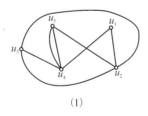
(1)
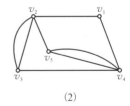
(2)

Fig. 1.15

4 There are n medicine boxes. Any two medicine boxes have the same kind of medicine inside and every kind of medicine is contained in just two medicine boxes. How many kinds of medicine are there?

5 There are n professors $A_1, A_2, \ldots, A_n$ in a conference. Prove that these n professors can be divided into two teams such that for every A_i, the number d_i of the people whom he has acquaintance with in another team is not less than d_i' in his team, $i = 1, 2, \ldots, n$.

6 There are 18 teams in a match. In every round, if one team competes with another team then it does not compete with the same team in another round. Now there have been 8 rounds. Prove that there must be three teams that have never competed with each other in the former 8 rounds.

7 n representatives attend a conference. For any four representatives, there is one person who has shaked hands with the other three. Prove that for any four representatives, there must be one person who shakes hands with the rest of the $n - 1$ representatives.

8 There are three middle schools, each of which has n students. Every student has acquaintance with $n + 1$ students in the other two schools. Prove that we can choose one student from each school such that the three students know each other.

9 There are $2n$ red squares on the a big chess board. For any two red squares, we can go from one of them to the other by moving horizontally or vertically to the adjacent red square in one step. Prove that all the red squares can be divided into n rectangules.

10 There are 2000 people in a tour group. For any four people, there is one person having acquaintance with the other three. What is the least number of people having acquaintance with all the other people in the tour group?

11 In a carriage, for any $m\,(m \geqslant 3)$ travelers, they have only one common friend. (If A is a friend of B, then B is a friend of A. Anyone is not a friend of himself.) How many people are there in the carriage?

12 There are five points A, B, C, D, E in the plane, where any three points are not on the same line. Suppose that we join some points with segments, called edges, to form a figure. If there are no above five points in the figure of which any three points are the vertices of a triangle in the figure, then there cannot be seven or more than seven edges.

Chapter 2 Degree of a Vertex

The *degree* of a vertex v in a graph G, denoted by $d_G(v)$, is the number of edges of G incident to v, where each loop is counted as two edges. Moreover, when there is no scope for ambiguity, we omit the letter G from graph-theoretic symbols and write, for example, $d(v)$ instead of $d_G(v)$. We denote by $\delta(G)$ and $\Delta(G)$ the minimum and maximum degrees of the vertices of G, or δ and Δ for brevity.

In Fig. 2.1, $d(v_1) = 1$, $d(v_2) = 3$, $d(v_3) = d(v_4) = 2$, $\delta = 1$, $\Delta = 3$.

A vertex is *odd* if its degree is odd, otherwise, it is *even*. In Fig. 2.1, v_1 and v_2 are odd vertices, and v_3 and v_4 are even.

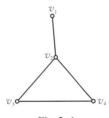
Fig. 2.1

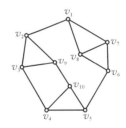
Fig. 2.2

A graph $G = (V, E)$ is said to be *k-regular*, if $d(v) = k$ for all $v \in V$. The complete graph on n vertices is $(n-1)$-regular. Fig. 2.2 shows a 3-regular graph.

The connection between the sum of the degrees of the vertices of a graph and the number of its edges is given as follows.

Theorem 1 For any graph G on n vertices, the sum of the degrees of all of the vertices is twice as large as the number of the edges. In symbols, if G with n edges has vertices $v_1, v_2, \ldots, v_n$, then

$$d(v_1)+d(v_2)+\cdots+d(v_n) = 2e.$$

Proof   The sum of the degrees of all of the vertices $d(v_1) + d(v_2)+\cdots+d(v_n)$ represents the whole number of the edges one of whose ends is $v_1, v_2, \ldots,$ or v_n. Since each edge has two ends, every edge of G is counted twice in the sum $d(v_1)+d(v_2)+\cdots+d(v_n)$. So the sum of the degrees of all of the vertices is twice as large as the number of the edges.

For instance, in Fig. 2.1, $e = 4$,

$$d(v_1)+d(v_2)+d(v_3)+d(v_4) = 1+3+2+2 = 8 = 2e.$$

Theorem 1 is often called the Hand-Shaking Lemma. A famous conclusion is given by Euler about two hundred years ago, that is to say, if many people shake hands when they meet, then the number of times of shaking hands is even. Then we can have the conclusion that there is an even number people who shake hands an odd number of times. The corollary is the following Theorem 2.

Theorem 2   In any graph G, the number of vertices with odd degree is even.

Proof   Suppose that G has vertices $v_1, v_2, \ldots, v_n$, where $v_1, \ldots, v_t$ are odd vertices and $v_{t+1}, \ldots, v_n$ are even. According to Theorem 1,

$$d(v_1)+\cdots+d(v_t)+d(v_{t+1})+\cdots+d(v_n) = 2e,$$

$$d(v_1)+\cdots+d(v_t) = 2e-d(v_{t+1})-\cdots-d(v_n).$$

Since $d(v_{t+1})+\cdots+d(v_n)$ are all even, the right side of the equality is even. However $d(v_1), \ldots, d(v_t)$ are all odd, then t must be even so that $d(v_1)+\cdots+d(v_t)$ is even. That is, the number of vertices with odd degrees is even.

Example 1   Among $n(n>2)$ people, there are at least 2 persons, where the number of their friends are the same.

Solution   We denote the n people by the vertices $v_1, v_2, \ldots, v_n$. If two persons are friends, we join the corresponding vertices. Then we get a graph. The assertion follows if we can find at least 2 vertices with the same degree in G.

A vertex is at most adjacent to other $n-1$ vertices in a simple graph on n vertices, so $d(v) \leqslant n-1$, for all $v \in V$. Hence the degree of a vertex in G can take only the following values:

$$0, 1, 2, \ldots, n-1.$$

However, not all of them are feasible. Note that a vertex with degree zero could not be adjacent to any other vertex and that the vertex with degree $n-1$ must be adjacent to any other $n-1$ vertices. So in G, only the following degrees are possible:

$$0, 1, 2, \ldots, n-2,$$

or

$$1, 2, 3, \ldots, n-1.$$

According to the pigeonhole principle, there are at least 2 vertices with the same degree.

Example 2 There are 24 pairs of contestants taking part in the International Table Tennis Mixed Doubles Contest. Some athletes shake hands before the game, and the two in one pair do not shake hands with each other. After the game, one male athlete asks everybody the number of hand-shaking, and all the answers are different. How many people does the male contestant's female partner shake hands with?

Solution The 48 vertices $v, v_0, v_1, \ldots, v_{46}$ represent the 48 contestants where the male contestant is represented by v, with edges joining two people who had shaken hands before, then we can get a graph G. In graph G, $d(v_i) \leqslant 46$, $i = 0, 1, 2, \ldots, 46$, and $d(v_i) \neq d(v_j)$, if $i \neq j$. So except v, the degree of the other vertices are

$$0, 1, 2, \ldots, 45, 46.$$

Without loss of generality, we suppose that $d(v_i) = i$, for $i = 0, 1, 2, \ldots, 46$. Vertex v_{46} is adjacent to every vertex except v_0, so v_0 and v_{46} are partners. Deleting v_{46}, v_0 and the edges which are incident to them, we can get graph G_1. The degree of every vertex in graph G_1 except v is still different and has decreased by 1. Likewise, v_{45} and v_1 are partners. So are v_{44} and v_2, ..., v_{24} and v_{22}. That is to say, the partner of v is v_{23}, so the male contestant's female partner shakes hands for 23 times.

Remark Changing 24 to 34 in Example 2, "male and female partners" to "group leader and deputy group leader", this is one of pre-selected problems in the 26th International Mathematics Olympiad. Changing 24 to 16, "male and female partners" to "2 football teams A and B", that is the third problem of the extra test of China Mathematical Competition in 1985.

Example 3 Every city in one country has 100 roads connecting to other cities, and any city can be reached from any other city. Now, one road is closed for repair. Prove that any city can still be reached from any other city.

Proof Assume that the road closed is AB. We need to prove that B can still be reached from A. Otherwise, except A, the degrees of all the vertices in the connected subgraph containing vertex A are even. By Theorem 2, it is a contradiction.

Remark The key point is to study the connected subgraph. The concept of connecting is very important, and we will encounter it later.

Example 4 Some cities are located on both sides of a river, and there are no less than 3 cities. Some routes connect the cities, and a pair of cities locating on both sides is connected by one route. Furthermore, each city is connected only with k cities on the other side. People could go to one city from any other city. Prove that people could still go to any city from any other city if one route is canceled.

Proof We may consider the two sides as the north side and the south side. The n cities on the north side are represented by x_1, x_2, ..., x_n and the whole set is denoted by $X = \{x_1, x_2, ..., x_n\}$. The m cities on the south side are represented by $y_1, y_2, ..., y_m$ and

$Y = \{y_1, y_2, \ldots, y_m\}$. If there is a route between the city x_i on the north side and the city y_i on the south side, we connect them to form an edge (x_i, y_i), and the set formed by all the edges is denoted by E. Then we can get a graph formed by the vertex sets X, Y and the edge sets E, which is called a *bipartite* graph, or an *even* graph, denoted by $G = (X, Y; E)$. The last two conditions are that there are only k edges incident to any vertex, and the graph G is connected in the sense that there exist some paths between any two vertices. The conclusion is that deleting any edge e from E, the graph is still connected.

Each vertex is incident to k edges, so

$$|X|k = |E| = |Y|k,$$

where $|X|$, $|E|$, $|Y|$ denote the numbers of the elements in sets X, E, Y respectively. So $|X| = |Y|$, and $n = m$. Since $|X| + |Y| \geq 3$, then $|X| = |Y| \geq 2$.

Deleting one edge from G, we can get a graph G'. If G' is not connected, G' is composed of two connected components G_1 and G_2.

Let

$$X = X_1 \cup X_2, X_1 \cap X_2 = \phi,$$
$$Y = Y_1 \cup Y_2, Y_1 \cap Y_2 = \phi,$$
$$G_1 = (X_1, Y_1; E_1), G_2 = (X_2, Y_2; E_2).$$

The deleted edge connects the vertices in X_1 and Y_2, then

$$|X_1|k - 1 = |E_1| = |Y_1|k,$$
$$|X_2|k - 1 = |E_2| = |Y_2|k - 1.$$

Then $(|X_1| - |Y_1|)k = 1$, and hence $k = 1$.

G is connected, so $|X| = 1$, which is contradictory to $|X| \geq 2$. So G' is connected, and the conclusion holds.

Example 5 There are 99 members in a club, and every member claims that he would play bridge only with someone he knows. Each member knows at least other 67 members. Prove that there must be 4 members who can play bridge together. (Polish Mathematics Contest in

1996)

Proof 1 Construct a graph G: denote the 99 members by 99 vertices and join every two vertices whose corresponding members know each other. The condition is that $d(v) \geqslant 67$, for all $v \in V$. We should prove that there is a complete graph K_4 in G. For any vertex u in G, $d(u) \geqslant 67$, so there exists a vertex v such that there are at most $(99-1-67=)31$ vertices which are neighbors of v, but not of u. Similarly, there are at most 31 vertices which are adjacent to u, not v. That is to say there are at least $(99-31-31-2=)35$ vertices adjacent to both u and v. As shown in Fig. 2.3, assume that vertex x is neighboring with both u and v. Vertex x is at least adjacent to one vertex y which is a neighbor of both u and v, since $d(x) \geqslant 67$, and there are at most $(31+31+2=)64$ vertices which are not adjacent to u and v at the same time. So u, v, x and y are neighbors, which proves the proposition.

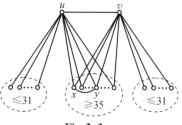

Fig. 2.3

Proof 2 The members are represented with the vertices, with edges joining two strangers, then we can get a graph G'. Since each person knows at least 67 people, $d(v) \leqslant 99-1-67=31$, for all v. We need to prove that there exist 4 vertices which are not mutually adjacent in G'. For a vertex u, choose a vertex v which is not adjacent to u. In the remaining 97 vertices, the number of vertices that are neighbors of either u or v could not exceed

$$d(u)+d(v) \leqslant 31+13=62.$$

So there exists a vertex x which is not adjacent to both u and v, and the number of vertices adjacent to u, v or x cannot exceed

$$d(u)+d(v)+d(x) \leqslant 31 \times 3=93.$$

Thus in the remaining 96 vertices, there must be a vertex y which is not adjacent to u, v and x, then the 4 people corresponding to u, v,

x and y know each other, and they can play bridge together.

Remark 1 Changing to 66, we might not be able to find out 4 people who know each other. The counter-example is shown in Fig. 2. 4. Separate the vertex set V into three subset $\{v_1, v_2, \ldots, v_{33}\}$, $\{v_{34}, v_{35}, \ldots, v_{66}\}$ and $\{v_{67}, v_{68}, \ldots, v_{99}\}$. Any two vertices in the same set are not adjacent, and two vertices in different sets are always adjacent. Obviously, the degree of every vertex is 66, and there are at least 2 vertices belonging to the same subset in any 4 vertices. It means that they are not adjacent. That is to say, there does not exist 4 vertices which are adjacent mutually.

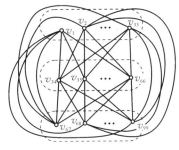

Fig. 2. 4

Remark 2 We can generalize it as follows: there are $n(n \geq 4)$ members in a club, and each person at least knows $\left[\dfrac{2n}{3}\right] + 1$ people, then there must exist 4 persons who know each other.

Remark 3 If G is a simple graph on n vertices, then we can get a graph by deleting all the edges belonging to G from the complete graph K_n, which is called the *complementary* graph of G, denoted by $\overline{G}$. The graph G in Proof 1 and the graph G' in Proof 2 are complementary.

Remark 4 We can make use of the pigeonhole principle to get another good solution. Please try it yourself.

Example 6 In a region, 20 members of a tennis club hold 14 single matches, and everybody plays at least once. Prove that there must be 6 matches in which 12 participants are all different. (USAMO 1989)

Proof The 20 members are represented by 20 vertices $v_1, v_2, \ldots, v_{20}$, with edges joining 2 participants who have played a match. Then we get a graph G. There are 14 edges in G, and denote the degree of the vertex by d_i, $i = 1, 2, \ldots, 20$, $d_i \geq 1$. According to Theorem 1,

$$d_1 + d_2 + \cdots + d_{20} = 2 \times 14 = 28.$$

Delete $d_i - 1$ edges incident to v_i. Since one edge might be deleted twice, the number of deleted edges does not exceed

$$(d_1 - 1) + (d_2 - 1) + \cdots + (d_{20} - 1) = 28 - 20 = 8.$$

So the graph G after deleting the edges, has at least $14 - 8 = 6$ edges, and the degree of each vertex in G' is at most 1. Therefore the 12 vertices that are incident to the 6 edges are different. That is to say, there must be 6 matches in which 12 participants are all different.

Example 7 Let $S = \{x_1, x_2, \ldots, x_n\}$ be a point set on the plane, and the distance between any 2 vertices is at least 1. Prove that there are at most $3n$ pairs of vertices, and the distance between 2 vertices in a pair is exactly 1.

Proof The n points are represented with n vertices, with edges joining 2 points whose distance is 1, then we get a graph G. The number of edges in G is denoted by e. Obviously, the vertex adjacent to vertex x_i in graph G is on the circle whose center is x_i, and whose radius is 1. Since the distance between two points in set S is no less than 1, there are at most 6 points of S on the circle. So $d(x_i) \leqslant 6$.

Using Theorem 1 to graph G, we have

$$d((x_1) + d(x_2) + \cdots + d(x_n) = 2e,$$

$$6n \geqslant 2e,$$

i.e. $e \leqslant 3n$. That is to say, the number of edges in graph G cannot exceed $3n$. So there are at most $3n$ pairs in the n vertices, whose distance of every pair is exactly 1.

Example 8 There are n points on the plane. Prove that the number of pairs of points whose distance is 1 would not exceed $\dfrac{n}{4} + \dfrac{\sqrt{2}}{2} n^{\frac{3}{2}}$.

Proof Denote the n points by n vertices in graph G. Let $V = \{v_1, v_2, \ldots, v_n\}$ is the vertex set in graph G. Join 2 vertices whose distance is 1. According to Theorem 1,

$$2e = d(v_1) + d(v_2) + \cdots + d(v_n).$$

C_i denotes a circle whose center is v_i and radius is 1. The total number of intersection points between any two of n circles does not exceed $2\binom{n}{2} = n(n-1)$. On the other hand, if v_i is adjacent to both v_j and v_k, then $v_i \in C_k \cap C_j$. Therefore, v_i regarded as an intersection point of circles $C_1, C_2, \ldots, C_n$ is counted $\binom{d(v_i)}{2}$ times, so

$$\binom{d(v_1)}{2} + \binom{d(v_2)}{2} + \cdots + \binom{d(v_n)}{2} \leqslant 2\binom{n}{2} = n(n-1). \quad \text{①}$$

According to the Cauchy Inequality, we have

$$\binom{d(v_1)}{2} + \binom{d(v_2)}{2} + \cdots + \binom{d(v_n)}{2} \geqslant \frac{2}{n}e^2 - e. \quad \text{②}$$

According to ① and ②, we have

$$\frac{2}{n}e^2 - e \leqslant n(n-1),$$

i.e.

$$2e^2 - ne - n^2(n-1) \leqslant 0.$$

Then

$$e \leqslant \frac{n}{4} + \frac{\sqrt{2}}{2}n^{\frac{3}{2}}.$$

Exercise 2

1 Let $G = (V, E)$ be a graph, where $|V| = n$ and $|E| = e$. Prove that $\delta \leqslant \dfrac{2e}{n} \leqslant \Delta$.

2 G is a graph with n vertices and $n+1$ edges. Prove that there exists at least one vertex whose degree is no less than 3.

3 Does there exists a polyhedron such that the number of its faces is odd and each face has an odd number of edges?

4 There are 15 telephones in a town. Can we connect them so that each telephone is connected with the other 5 telephones?

5 There are 123 people attending an academic symposium, and each person has discussed with at least 5 other participants. Prove that there is at least one person who has discussed with more than 5 participants.

6 In a conference, each councilor does not know at most 3 person. Prove that all the councilors can be divided into 2 groups such that each councilor in a group does not know at most 1 person.

7 $2n$ people are getting together, and each one knows at least n people. Prove that there must exist 4 persons such that they know the persons sitting besides them when sitting around a round table ($n \geqslant 2$).

8 There are 9 people $v_1, v_2, \ldots, v_9$. v_1 has shaken hands with 2 people. v_2 and v_3 have shaken hands with 4 people respectively. v_4, v_5, v_6 and v_7 have shaken hands with 5 people respectively. v_8 and v_9 have shaken hands with 6 people respectively. Prove that there must exist 3 people who have shaken hands with each other among themselves.

9 There are 14 members in a tour group. When they are resting on a hill, they would like to play bridge, and each of them has all cooperated with 5 of them before. There is a rule that 4 people can play together only when any 2 of them have never cooperated with each other. If so, they cannot go on after 3 rounds. At this time, there comes another tourist, and he has never cooperated with the members in the tour group. Prove that there must be another round if the new comer joins the bridge.

10 For a vertex set P consists of any n vertices on the plane, the vertex set is stable if the distance between any 2 of them is fixed. Prove that the vertex set P consists of $n(n \geqslant 4)$ is stable if any 3 of them are not on the same line, and there are $\frac{1}{2}n(n-3)+4$ pairs whose distance are determined.

11 A cube with edge of length n is cut into n^3 unit cubes by planes parallel to its sides. How many pairs of unit cubes whose common vertices are no more than 2 are there?

12 There are 21 routes connecting the capital and other cities. There is only one route connecting city A and one other city, and all cities other than A are connected with some other places. Prove that city A can be reached from the capital.

Chapter 3　　　　　Turán's Theorem

In 1941, a Hungarian mathematician Turán brought forward his famous theory so as to answer the question that if a graph with n vertices does not contain a complete graph K_m with m vertices as its subgraph, how many edges can the graph contain at most? Then a new branch of graph theory called "extremal graph theory" appeared. The extremal graph theory is one of the most active branch of graph theory. In 1978, a Hungarian mathematician B. Bollobás wrote a book called "extremal graph theory" which is the authoritative book of this branch.

Then we begin with the definition of a k-partite graph.

If the set of vertices V of a graph G can be decomposed into a union set of k subsets which do not intersect each other. Namely,

$$V = \bigcup_{i=1}^{k} V_i, \ V_i \cap V_j = \varnothing, \ i \neq j,$$

and there is no edge whose two vertices are in the same subset. We call such graph a k-partite graph denoted by $G = (V_1, V_2, \ldots, V_k; E)$.

Fig. 3.1 shows us a 2-partite graph which is also called *bigraph*. Fig. 3.2 shows us a 3-partite graph.

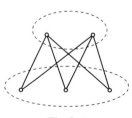

Fig. 3.1

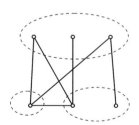

Fig. 3.2

Clearly, any graph with n vertices is n-partite graph.

Suppose that there is a k-partite graph $G = (V_1, V_2, \ldots, V_k; E)$ with $|V_i| = m_i$. A graph G is said to be a complete k-partite graph if any two vertices of G satisfy $u \in V_i$, $V \in V_j$, $i \neq j$, where $i, j = 1, 2, \ldots, k$, u and v are adjacent. We denote G by $K_{m_1, m_2, \ldots, m_k}$. Fig. 3.1 shows us a complete bigraph $K_{2,3}$. There are m^2 and $m(m+1)$ edges in complete bigraphs $K_{m,m}$ and $K_{m,m+1}$, so the number of edges of the graph is $\left[\dfrac{n^2}{4}\right]$, where n is the number of vertices in G. (Here $[x]$ denotes the largest integer no more than x.) Complete bigraphs $K_{m,m}$ and $K_{m,m+1}$ contain no triangle. In Theorem 1, we can see that these two kinds of graphs contain the most number of edges among the graphs without triangles.

Theorem 1 If a graph G with n vertices contain no triangle, the largest number of edges of G is $\left[\dfrac{n^2}{4}\right]$.

Proof Assume that v_1 is the vertex with the maximum degree in G, $d(v_1) = d$ and we denote d vertices adjacent to v_1 by

$$v_n, v_{n-1}, \ldots, v_{n-d+1}.$$

Since G contains no triangle, any two of $v_n, v_{n-1}, \ldots, v_{n-d+1}$ are not adjacent. So the number of edges of G satisfies

$$e \leqslant d(v_1) + d(v_2) + \cdots + d(v_{n-d})$$
$$\leqslant (n-d) \cdot d \leqslant \left(\frac{n-d+d}{2}\right)^2$$
$$= \frac{n^2}{4}.$$

Since e must be an integer, $e \leqslant \left[\dfrac{n^2}{4}\right]$.

The upper bound can be met only when $G = K_{m,m}$ if $n = 2m$ and $G = K_{m,m+1}$ if $n = 2m+1$.

We can also use induction to prove the theorem, and we leave it as an exercise.

Example 1 Suppose that there are 20 vertices and 101 edges in a graph G. Prove that there must exist two triangles which have a common edge in G.

Proof In general, we can replace 20 by more general number $2n$ ($n \geq 2$). Now we use induction, if there are $2n$ ($n \geq 2$) vertices and $n^2 + 1$ edges in a graph G, there must be two triangles which have a common edge in G.

When $n = 2$, there are 4 vertices and 5 edges in the graph G. Consider the complete graph K_4, there are $\binom{4}{2}$ edges in K_4. It is not difficult to prove that whatever edge we remove from K_4, there must still be two triangles which have a common edge in G. So the theorem holds when $n = 2$.

Suppose that the theorem holds for $n = k$ ($k = 2$). Let G be a graph with $2(k+1)$ vertices which are denoted by $v_1, v_2, \ldots, v_{2k+2}$ and $(k+1)^2 + 1 = k^2 + 2k + 2$ edges. Note that

$$\left[\frac{(2k+2)^2}{4}\right] = [k^2 + 2k + 1] < k^2 + 2k + 2.$$

According to Theorem 1, there must be a triangle in G. Without loss of generality, we denote this triangle by $\triangle v_1 v_2 v_3$ and $d(v_1) \leq d(v_2) \leq d(v_3)$.

If one of the vertices $v_4, v_5, \ldots, v_{2k+2}$ is adjacent to two of v_1, v_2, v_3, then we get two triangles with a common edge.

If each of $v_4, v_5, \ldots, v_{2k+2}$ is adjacent to at most one of v_1, v_2, v_3, then the number of edges which join the vertex set $\{v_4, v_5, \ldots, v_{2k+2}\}$ to $\{v_1, v_2, v_3\}$ is less than $(2k+2) - 3 = 2k - 1$. Then the number of edges which join the vertex set $\{v_1, v_2\}$ to $\{v_4, v_5, \ldots, v_{2k+2}\}$ is no more than $\frac{2}{3}(2k-1)$. We remove the vertices v_1, v_2 and the edges adjacent to them from G to get G'. The number of vertices in G' is $2k$ and the number of edges is

$$e' \geq k^2 + 2k + 2 - 3 - \frac{2}{3}(2k-1)$$

$$= k^2 + \frac{2}{3}k - \frac{1}{3} \geq k^2 + 1, \text{ since } k \geq 2.$$

By induction, there are two triangles with a common edge in G', which are also in G. Then we complete the proof.

Example 2 We denote integer pair by (a, b) $(1 \leqslant a, b \leqslant n, a \neq b)$ by S, where (a, b) is the same as (b, a). Prove that there are at least $\dfrac{4m}{3n}\left(m - \dfrac{n^2}{4}\right)$ triples (a, b, c) satisfying (a, b), (a, c) and $(b, c) \in S$. (Asian-Pacific regional Mathematical Olympiad in 1989)

Solution We construct a graph G and denote the number i by v_i, where $i = 1, 2, \ldots, n$, and $(i, j) \in S$. The vertex v_i is adjacent to v_j if and only if $(i, j) \in S$. Then there are n vertices and m edges in G. What we will prove is that there are at least $\dfrac{4m}{3n}\left(m - \dfrac{n^2}{4}\right)$ triangles in G.

Let the degree of vertex v_i be d_i and denote the edge set of G by E. If $(v_i, v_j) \in E$, then there are $d_i + d_j - 2$ edges joining v_i, v_j with all other $n - 2$ vertices. So there are at least $d_i + d_j - n$ pairs of edges joining v_i, v_j with the same vertex. These edges together with the edge (v_i, v_j) form triangles. So there are at least $d_i + d_j - n$ triangles containing (v_i, v_j) in G. Since we have counted every triangle containing the edge (v_i, v_j) in G three times, there are

$$k = \frac{1}{3} \sum_{(v_i, v_j) \in E} (d_i + d_j - n)$$

triangles in G. Since the degree d_i of vertex v_i has been counted 3 times in the above equation and the number of edges is m. So

$$K = \frac{1}{3}\left(\sum_{i=1}^{n} d_i^2 - mn\right) \qquad ①$$

Since $\sum_{i=1}^{n} d_i = 2m$, apply the Cauchy Inequality to ① and get

$$k \geqslant \frac{1}{3}\left[\frac{1}{n}\left(\sum_{i=1}^{n} d_i\right)^2 - mn\right]$$

$$= \frac{1}{3}\left(\frac{4m^2}{n} - mn\right)$$

$$= \frac{4m}{3n}\left(m - \frac{n^2}{4}\right).$$

Remark The problem derives from a question in graph theory: if there are m edges and n vertices in a graph G, there must be no less than $\frac{4m}{3n}\left(m - \frac{n^2}{4}\right)$ triangles in the graph G.

Suppose that $n = mk + r$ ($k \geq 1$, $0 \leq r < m$). We denote the complete m-partite graph $K_{n_1, n_2, \ldots, n_m}$ by $T_m(n)$, where $n_1 = n_2 = \cdots = n_r = k+1$, $n_{r+1} = \cdots = n_m = k$. We also denote the number of edges of $T_m(n)$ by $e_m(n)$. Fig. 3.3 shows us $T_3(5)$, where $e_3(5) = 8$. The formula to calculate $e_m(n)$ is:

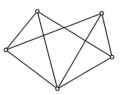

Fig. 3.3

$$e_m(n) = \binom{n-k}{2} + (m-1)\binom{k+1}{2}, \quad k = \left[\frac{n}{m}\right].$$

We leave it as an exercise.

Let $G = (V_1, V_2, \ldots, V_m; E)$ be an m-partite graph G with n vertices, and $p_i = |v_i|$ ($\sum_{i=1}^{m} p_i = n$). We can show that the number of edges in G is less than $e_m(n)$ and the equality holds if and only if $G \cong T_m(n)$. (We leave the proof as an exercise.) In other words, $T_m(n)$ is the only m-partite graph with n vertices, which has the most number of edges.

Clearly, any m-partite graph contains no K_{m+1}. Furthermore, Turán prove that $T_m(n)$ is the only m-partite graph with n vertices, which contains the most number of edges and no K_{m+1}.

Theorem 2 Suppose G contains no K_{m+1}, then $e(G) \leq e_m(n)$. The equality holds if and only if $G \cong T_m(n)$.

This is Turán's theorem and we omit the proof. If you are interested in it, you can read *Graph Theory and Its Applications* written by J. A. Bondy and U. S. A. Murty.

Example 3 Let $A_1, A_2, A_3, A_4, A_5, A_6$ be six points on a plane and there are no three points on a line.

(1) If we join the points randomly and get 13 line segments. Prove that there must exist four points so that each of them is adjacent to any other three points.

(2) If there are only 12 line segments joining these points. Draw a graph to show that the conclusion of (i) is not true.

(3) Can the conclusion of (i) be modified so that there must exist four copies of K_4? Give a counterexample or prove it.

Solution

(1) We can phrase the problem in the language of graph theory: There are 6 vertices and 13 edges in a graph G, prove that the G contains K_4.

It is easy to calculate $e_4(6) = 12 < 13$. According to Theorem 2, we know that G must contain K_4.

(2) Consider the complete 3-partite graph $K_{2,2,2}$. According to Fig. 3.4, we choose any 4 vertices from $K_{2,2,2}$ and there must be 2 vertices belonging to one part. These two vertices are not adjacent. So the 4 vertices we choose arbitrarily cannot form a K_4.

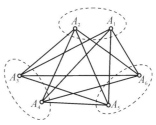

Fig. 3.4

Remark (1) Of course, we can use Theorem 2 to prove it and there are many other ways. Here we list two different methods.

(1) Since the sum of the degrees of 6 vertices is $2 \times 13 = 26$, there are at least 2 vertices whose degrees are 5 among the 6 vertices. Otherwise, the sum of degrees is $5 + 5 \times 4 = 25 < 26$. Without loss of generality, suppose that $d(A_1) = d(A_2) = 5$, there are 9 edges incident to A_1 or A_2. According to Fig. 3.5, there are $13 - 9 = 4$ edges joining A_3, A_4, A_5, A_6. Two ends of any of the four edges together with A_1, A_2 can form a K_4.

(2) Since there are 15 edges in a complete graph with 6 vertices, we delete two edges. We discuss the problem in two different cases.

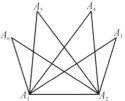

Fig. 3.5

① If there are common vertices in two deleted edges as Fig. 3.6 shows us, then A_2, A_4, A_5, A_6 form a K_4.

② If there is no common vertex in two deleted edges as Fig. 3.7 shows us, then A_1, A_3, A_5, A_6 form a K_4.

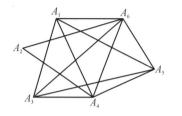 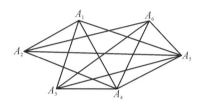

Fig. 3.6   Fig. 3.7

(3) According to the above two cases, let us begin our discussion:
In case ①, there are six K_4: (A_1, A_4, A_5, A_6), (A_2, A_4, A_5, A_6), (A_3, A_4, A_5, A_6), (A_2, A_3, A_4, A_5), (A_1, A_3, A_4, A_5), (A_1, A_3, A_5, A_6).

In case ②, there are four K_4: (A_1, A_3, A_5, A_6), (A_1, A_4, A_5, A_6), (A_1, A_3, A_4, A_5), (A_1, A_2, A_4, A_5). So there must exist four copies of K_4.

Example 4 In a simple graph with eight vertices, can you find the maximum number of edges of a graph which contains no quadrangle? (The quadrangle consists of four vertices A, B, C, D and four edges AB, BC, CD, DA.) (China Mathematical Olympiad in 1992)

Solution The maximum number of edges is 11.

First, Fig. 3.8 shows us a graph with 8 vertices and 11 edges, which contains no quadrangle.

Next, we will prove the fact that if any simple graph contains 12 edges, the graph must contain a quadrangle.

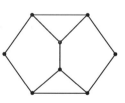

Fig. 3.8

First, we must point out two obvious facts.

(a) Let A, B be two vertices ($A \neq B$). If A is adjacent to vertices $C_1, C_2, \ldots, C_k$ and B is adjacent to at least two of vertices $\{C_1, \ldots,$

C_k}, the graph must contain a quadrangle.

(b) If there are 5 edges joining 4 vertices, the graph must contain a quadrangle.

Suppose that a graph contains 8 vertices and 12 edges, which contains a quadrangle. A is one of the vertex which is incident to the most vertices.

(1) Suppose that A is incident to $s \geq 5$ edges and we denote the edge set incident to A by S and the vertex set other than A and S by T. According to (a) and (b), we know that the number of the edges joining the vertices in S is no more than $\left[\frac{s}{2}\right]$. The number of the edges joining the vertices in T is at most $\binom{|T|}{2}$. The number of edges joining the vertices in S and T is at most $|T|$. So the sum of edges is at most

$$s + \left[\frac{s}{2}\right] + \binom{|T|}{2} = 7 + \left[\frac{s}{2}\right] + \binom{|T|}{2}.$$

When $s \geq 5$, the number of edges is less than 12. A contradiction.

(2) There are four edges from A, namely: $AA_j (j = 1, 2, 3, 4)$. Let another three vertices be B_1, B_2, B_3. There are at most two edges joining {A_1, A_2, A_3, A_4} and at most three edges joining {B_1, B_2, B_3}. There are at most three edges joining these two vertex sets. Since there are 12 edges in the graph G, the number of edges in each of the three groups is 2, 3, 3 respectively. Without loss of generality, let the three edges in the third group be $A_j B_j (j = 1, 2, 3)$. Since there are two edges without a common vertex in the first group, there is one edge joining A_1, A_2, A_3, denoted by $A_1 A_2$. So $A_1 A_2 B_2 B_1$ is a quadrangle. A contradiction.

(3) There are three edges with an end point at A and every vertex is incident to three edges. Suppose that A, B are not adjacent. We denote the three edges with an endpoint at A or B by AA_j, $BB_j (j = 1, 2, 3)$, respectively. So according to (a), there are at most one

common vertex between $\{A_1, A_2, A_3\}$ and $\{B_1, B_2, B_3\}$.

If there is no common vertex between them, there is at most one edge joining two of the three vertices. There are at most three edges joining the two three-vertex sets, the number of edges is at most 11. It is a contradiction.

If there are only one common vertex between the two three-vertex set, we consider the eighth vertex. According to the Pigeonhole Principle, among the three edges having the common vertex as an end there must be two edges which have ends in one three-vertex set. Then we get a quadrangle. It is a contradiction.

In conclusion, we have proved that the graph containing 8 vertices and 12 edges must contain a quadrangle. So the maximum number of edges is 11.

Example 5 G is a simple graph with n vertices. If G contains no quadrangle, then the number of edges is

$$e \leqslant \frac{1}{4} n (1 + \sqrt{4n - 3}).$$

Solution We denote the vertex set of G by $V = \{v_1, v_2, \ldots, v_n\}$. For any vertex $v_i \in V$, the number of vertex pairs (x, y) adjacent to v_i is $\binom{d(v_i)}{2}$. Because G contains no quadrangle, when v_i is changing in the V, all the vertices pairs (x, y) are distinct. Otherwise, vertices pairs (x, y) are counted in both $\binom{d(v_i)}{2}$ and $\binom{d(v_j)}{2}$, respectively. Then v_i, x, v_j, y construct a quadrangle. So

$$\sum_{i=1}^{n} \binom{d(v_i)}{2} \leqslant \binom{n}{2}.$$

In virtue of the Cauchy inequality,

$$\sum_{i=1}^{n} \binom{d(v_i)}{2} = \frac{1}{2} \sum_{i=1}^{n} d^2(v_i) - e$$

$$\leqslant \frac{1}{2} \cdot \frac{1}{n} \Big(\sum_{i=1}^{n} d(v_i)\Big)^2 - e$$
$$= \frac{2}{n} e^2 - e.$$

So
$$\frac{2}{n} e^2 - e \leqslant \binom{n}{2},$$
$$e^2 - \frac{n}{2} e - \frac{1}{4} n^2 (n-1) \leqslant 0.$$

We solve the above inequality and get
$$e \leqslant \frac{n}{4}(1 + \sqrt{4n-3}).$$

Remark This problem tells us an upper bound of the number of edges of graphs which contain n vertices and no quadrangle. But it is not the maximum number. For general n, we can do a further research into its maximum number. Example 4 has showed when $n = 8$, the maximum number is 11.

Example 6 There are n vertices and l edges in a graph. Then $n = q^2 + q + 1$, $l \geqslant \frac{1}{2} q(q+1)^2 + 1$, $q \geqslant 2$, $q \in \mathbf{N}$.

We know that any four points in the graph do not lie on one plane and every point must lie on at least one line. So there exists a point that lies on at least $q + 2$ lines. Prove that the graph must contain a quadrangle in the space, consisting of four points A, B, C, D and four lines AB, BC, CD, DA. (China Mathematical Competition in 2003)

Solution The condition that any four points cannot lie on a plane is to ensure that there are no three points on a line. So in terms of graph theory, we only need to prove that the graph contains a quadrangle. To solve this problem, we need to use the idea of Example 5, but we cannot use it directly. Consider the removal of the $d(v_1)$ vertices which are adjacent to v_1 ($d(v_1) \geqslant q + 2$). There will be $\binom{n - d(v_1)}{2}$ pairs of vertices left.

As in Example 5, when the graph contains no quadrangle,

$$\binom{n-d(v)}{2} \geq \sum_{i=2}^{n} \binom{d(v_i)-1}{2}.$$

Similarly,

$$\sum_{i=2}^{n} \binom{d(v_i)-1}{2} = 2l - n + 1 - d(v).$$

Then use the Cauchy equality,

$$\frac{[n-d(v_1)][n-d(v_1)-1]}{2}$$

$$\geq \frac{1}{2}\left\{\sum_{i=2}^{n}[n-d(v_i)]^2 - \sum_{i=2}^{n}[n-d(v_i)]\right\}$$

$$\geq \frac{1}{2}\left\{\frac{1}{n-1}[2l-n+1-d(v_1)]^2 - [2l-n+1-d(v_i)]\right\}.$$

That is to say,

$$(n-1)[n-d(v_1)][n-d(v_1)-1]$$
$$\geq [2l-n+1-d(v_1)][2l-2n+2-d(v_1)]$$
$$\geq [q^3+q^2-d(v_1)+2][q^3-q+2-d(v_1)]$$
$$= [nq-q+2-d(v_1)][nq-q-n+3-d(v_1)].$$

It contradicts the fact that

$$(q+1)[n-d(v_1)] < nq - q + 2 - d(v_1)$$

and

$$q[n-d(v_1)-1] \leq nq - 1 - n + 3 - d(v).$$

So the graph must contain a quadrangle.

As an application of Turán's Theorem, we give an example in geometry.

In a point set S on the plane, let the maximum distance of any two points be d. If d is a finite number, we call d the diameter of the vertex set S.

Let $S = \{x_1, x_2, \ldots, x_n\}$ be a point set which consists of n

points and whose diameter is 1. The n points determine the distance of two points in the $\binom{n}{2}$ point pairs. For a number d between 0 and 1, we can ask the following question: in a vertex set $S = \{x_1, x_2, \ldots, x_n\}$ whose diameter is 1, how many points pairs are there such that the distance between two points in the point pairs is more than d? Here, we discuss only the special case when $d = \frac{\sqrt{2}}{2}$.

First, when $n = 6$, then $S = \{x_1, x_2, x_3, x_4, x_5, x_6\}$. We put them on the vertices of a regular hexagon so that the distance of two points in the point pairs (x_1, x_4), (x_2, x_5), (x_3, x_6) is 1. Fig. 3.9 shows us that the diameter of S is 1. It is not difficult to find that the distance of two points in the point pairs (x_1, x_3), (x_2, x_4), (x_3, x_5), (x_4, x_6), (x_5, x_1), (x_6, x_2) is $\frac{\sqrt{3}}{2}$. So there are 9 point pairs in the point set S whose diameter is 1. The distance of these 9 points pairs is more than $\frac{\sqrt{2}}{2}$.

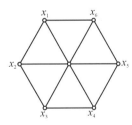 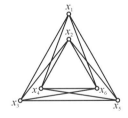

Fig. 3.9　　　　　　　Fig. 3.10

But 9 is not the best answer for 6 points. Suppose we arrange the 6 points as Fig. 3.10 shows us, namely, the vertices x_1, x_3, x_5 form a regular triangle with edges of length 1. The vertices x_2, x_4, x_6 construct a regular triangle the length of whose edges of length 0.8. The center of the new triangle coincide with the center of $\triangle x_1 x_3 x_5$ and the edges of the new triangle are paralleled to those of $\triangle x_1 x_3 x_5$, then the distance between two points in the point pairs other than $(x_1,$

x_2), (x_3, x_4), (x_5, x_6) is more than $\frac{\sqrt{2}}{2}$. So we have $\binom{6}{2} - 3 = 12$ point pairs and the distance between two points in the point pairs is more than $\frac{\sqrt{2}}{2}$. In fact, it is the best answer we can get. For the general case, Theorem 3 gives the a solution to the question.

Theorem 3 Let $S = \{x_1, x_2, \ldots, x_n\}$ be a point set on the plane whose diameter is 1, then the maximum possible number of the point pairs the distance of which is more than $\frac{\sqrt{2}}{2}$ is $\left[\frac{n^3}{3}\right]$. For every n, there exists a point set $\{x_1, x_2, \ldots, x_n\}$ whose diameter is 1, and there are exactly $\left[\frac{n^3}{3}\right]$ point pairs such that the distance of two points in each pair is more than $\frac{\sqrt{2}}{2}$.

Proof Draw a graph G: we denote n points by n vertices. Two vertices are adjacent if and only if the distance of two vertices is more than $\frac{\sqrt{2}}{2}$. First, we prove G contain no K_4.

For any four points on the plane, their convex hull could have only three case: a line segment, a triangle, or a quadrangle, as Fig. 3.11 shows us. Clearly in every case there is an angle $\angle x_i x_j x_k$ no more than $90°$. For the three vertices x_i, x_j, x_k, it is impossible that the distance of any two vertices of the three points is all greater than $\frac{\sqrt{2}}{2}$ and less than or equal to 1. Here, we denote the distance between

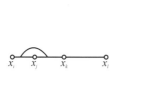
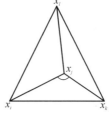

Fig. 3.11

x and y by $d(x, y)$, $d(x_j, x_k)$ is more than $\frac{\sqrt{2}}{2}$ for all j, k and $\angle x_i x_j x_k \geq 90°$, then

$$d(x_i, x_k) \geq \sqrt{d^2(x_i, x_j) + d^2(x_j, x_k)} > 1.$$

Since the diameter of the vertex set S is 1, among any four vertices in G there is at least one pair whose vertices are not adjacent. It means that G contains no K_4.

According to Theorem 2 the number of edges of G is no more than $e_3(n) = \left[\frac{n^3}{3}\right]$.

We can construct a vertex set $\{x_1, x_2, \ldots, x_n\}$ which contains $\left[\frac{n^3}{3}\right]$ vertex pairs so that the distance of two vertices in each pair is more than $\frac{\sqrt{2}}{2}$. The construction is as follows. Choose r so that $0 < r < \frac{1}{4}\left(1 - \frac{\sqrt{2}}{2}\right)$. Then draw three circles whose radii are all 1 and the distance of any two of their centers is all $1 - 2r$. As Fig. 3.12 shows us, we put $x_1, x_2, \ldots, x_{\left[\frac{n}{3}\right]}$ in a circle, $x_{\left[\frac{n}{3}\right]+1}, \ldots, x_{\left[\frac{2n}{3}\right]}$ in another circle and $x_{\left[\frac{2n}{3}\right]+1}, \ldots, x_n$ in the third circle so that the distance of x_1 and x_n is 1. Obviously, the diameter of this set is 1. Furthermore, $d(x_i, x_j) > \frac{\sqrt{2}}{2}$ if and only if x_i and x_j belong to different circles. So there exist exactly $\left[\frac{n^3}{3}\right]$ vertex pairs (x_i, x_j) such that $d(x_i, x_j) > \frac{\sqrt{2}}{2}$.

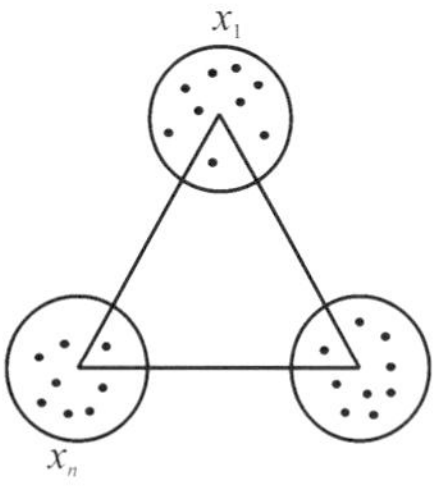

Fig. 3.12

Exercise 3

1 Prove that if a bigraph $G = (X, Y; E)$ is δ-regular, then

$|X| = |Y|$.

2 Prove by induction Theorem 1.

3 Draw a simple graph which contains 20 vertices, 100 edges and no triangle.

4 Prove that if there are $2n+1$ vertices and n^2+n+1 edges in a simple graph G, then G must contain a triangle.

5 (1) The number of edges in a complete m-partite graph with n vertices $e_m(n) = \binom{n-k}{2} + (m-1)\binom{k+1}{2}$, where $k = \left[\frac{n}{m}\right]$.

(2) Let G be a complete m-partite graph with n vertices. Then $e(G) \leqslant e_m(n)$.

6 There are n students from each of the two countries X and Y. Every student from X has danced with some (but not all) students from Y and every student from Y has danced with at least one student from X. Prove that it is possible to find two students x, x' from X and two students y, y' from Y so that x has danced with y and has not danced with y' while x' has danced with y' and has not danced with y.

7 Use Turán's Theorem to prove Problem 9 in Exercise 2.

8 If a graph G contains n $(n > 5)$ vertices, then G and its complement $\overline{G}$ contain at least $\frac{1}{24}n(n-1)(n-5)$ triangles in all.

9 X is a set of n elements and set its m k-subsets $A_1, A_2, \ldots, A_m$ red k-subset. Prove that if $m > \frac{(k-1)(n-k)+k}{k^2} \cdot \binom{n}{k-1}$, then there must exist a $(k+1)$-subset of X so that all k-subsets are red k-subset.

10 Let $K_{3,3}$ be a graph. Prove that a graph with 10 vertices and 40 edges must contain a $K_{3,3}$.

11 In a circular city whose radius is 6 kilometers, there are 18 police cars making the rounds. They use wireless equipments to communicate with each other. If the wireless is effective within 9 kilometers. Prove that at any time, there must exist two cars, each of them can communicate with five other cars.

12 There are $2n$ points in the space and any four of them are not on one plane. If there are $n^2 + 1$ line segments joining them, then these line segments can form at least n distinct triangles.

Chapter 4   Tree

Among all kinds of graphs, there is a simple but important graph, which is called "tree". Tree is very important because it is widely used not only in many other fields but also in graph theory itself. In graph theory, tree is a very simple graph, so when we discuss some general graph theory conjectures, we may study tree first.

First, we introduce several concepts.

Suppose, in graph G, a sequence consists of different edges:

$$e_1, e_2, \ldots, e_m.$$

If edge $e_i = (v_{i-1}, v_i)$, $i = 1, 2, \ldots, m$, then we call the sequence a *chain* from v_0 to v_m. The number m is called the *length* of the chain. v_0 and v_m are its ends. We denote the chain by $v_0 v_1 \ldots v_m$.

A path is a chain where the v_i are all distinct

If the ends v_0 and v_m of a chain are the same, then the chain is called a *cycle*①.

In Fig. 4.1, e_1, e_2, e_3, e_4, e_5 form a chain, whereas e_1, e_2, e_3 constitute a cycle. If for any two vertices u and v of graph G, there is a chain from u to v, then G is called a *connected* graph. Otherwise, G is called a disconnected graph.

The graph in Fig. 4.1 is a connected graph. The graph in Fig. 4.2 is a disconnected graph.

Now, let us give the definition of a tree.

---

① In this book, "cycle" is actually a "closed chain". It is different from the "cycle" in general graph theory. The "cycle" in graph theory is a "closed chain" $v_0 v_1 \ldots v_m (v_0 = v_m)$, in which vertices $v_1, v_2, \ldots, v_m$ are distinct.

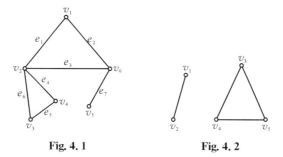

Fig. 4.1   **Fig. 4.2**

A connected graph which contains no cycle is called a *tree*. We usually denote a tree by T.

According to the definition of tree, tree is obviously a simple graph. A tree with eight vertices is shown in Fig. 4.3. Clearly, a graph without cycles must be composed of one or several trees whose vertices are disjoint. We call such graph a *forest*.

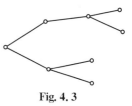

Fig. 4.3

The graph in Fig. 4.4 is a forest, which is composed of three trees. The vertex with degree 1 is called a *pendant* vertex (or leaf).

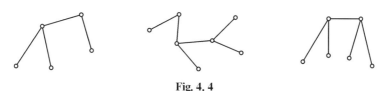

Fig. 4.4

Theorem 1   If a tree T has no less than 2 vertices, then T contains at least two pendant vertices.

Proof 1   Suppose we start from some vertex u, along the edges of T, every edge can only be passed by once. Since a tree has no cycle, it cannot return to the vertices which have been passed. It means that each vertex can be passed at most once. If the vertex we pass is not a pendant vertex, because its degree is more than 1, we can continue. But the number of vertices of T is finite, so it is impossible to continue forever. If we cannot continue further at v, then v is a pendant vertex.

We start from a pendant vertex v and we can end up at another pendant vertex v'. So tree T has at least two pendant vertices.

Proof 2 Assume that $\mu = uv_1v_2\cdots v_kv$ is the longest chain of T, we can show that $d(u) = d(v) = 1$, i.e. u, v are pendant vertices.

In fact, if $d(u) \geq 2$, then there exists vertex w ($w \neq v_1$) adjacent to u. If w is one of $v_2, \ldots, v_k, v$, then a cycle occurs. It contradicts the definition of tree. If w is different from $v_2, \ldots, v_k$, v, then $wuv_1\cdots v_kv$ is a chain longer than μ, a contradiction. So $d(u) = 1$. Similarly, $d(v) = 1$. Hence tree T has at least two pendant vertices.

Remark Proof 1 is "by construction" and Proof 2 is by the method of "the longest chain". They are two very important methods.

Theorem 2 Let n be the number of vertices of T, then the number of edges $e = n - 1$.

Proof We prove by induction on n.

When $n = 1$, $e = 0$, the conclusion holds. Suppose the conclusion holds for $n = k$. Let T be a tree on $k + 1$ vertices ($k \geq 1$). By Theorem 1, T has at least two pendant vertices. Let v be one of them, then deleting v and its adjacent edges. We can get tree T' on k vertices. According to the induction hypothesis, T' has $k - 1$ edges. So the number of edges of T is k, hence the conclusion holds for any natural number n.

Theorem 3 Let T be a graph with n vertices and e edges. Then the following three propositions are equivalent:

(1) T is a tree;

(2) T has no cycle, and $e = n - 1$;

(3) T is connected, and $e = n - 1$.

Proof (1)→(2):

Suppose T is a tree. According to the definition of T, T has no cycle. By Theorem 2, $e = n - 1$. Hence (2) holds.

(2)→(3):

It suffices to show that T is connected and T is disconnected.

Suppose T has k connected branches. Since each connected branch has no cycle, hence each connected branch is a tree. If the ith branch has p_i vertices, according to Theorem 2, the ith branch has $p_i - 1$ edges. So

$$e = (p_1 - 1) + \cdots + (p_i - 1) = n - k \leqslant n - 2.$$

This contradicts $e = n - 1$. Hence T is connected.

(3)→(1):

It suffices to show that T has no cycle, then T is a tree. When $n = 1$, the conclusion holds. Suppose $n \geqslant 2$, then T must have a pendant vertex. Otherwise, since T is connected and $n \geqslant 2$, the degree of each vertex of T is no less than 2. Hence

$$e = \frac{1}{2}[d(v_1) + d(v_2) + \cdots + d(v_n)] \geqslant \frac{1}{2} \times 2n = n.$$

This contradicts $e = n - 1$.

Now by induction on n, we will show that T has no cycle.

When $n = 2$, $e = 1$, T is has no cycle.

Suppose the proposition holds for $n = k$. T is a graph on $k + 1$ vertices, v is a pendant vertex. Deleting v and its adjacent edges, we obtain graph T'. By induction and T' has no cycle. Add v and its adjacent edges and we can obtain graph T again. Hence the proposition is correct.

Theorem 3 shows that any two conditions of "connected", "having no cycle", "$e = n - 1$" imply that T is tree. Hence they all can be the definition of a tree.

Example 1 If T is a tree, then

(1) T is connected, but after deleting any edge of T, the obtained graph G is disconnected;

(2) T has no cycle, but after adding any edge, the obtained graph G contains only one unique cycle.

Conversely, if T satisfies (1) and (2), then T is tree.

Proof (1) If graph G is connected, then G is still a tree, so G has $n - 1$ edges, equal to the number of edges of T, a contradiction.

(2) If G has no cycle, then G is still a tree. So G has $n-1$ edges, equal to the number of edges of T. It is impossible, so G contains a cycle. Clearly, G contains only one cycle.

If (1) and (2) hold, then T is tree. We leave it as an exercise.

This exercise characterizes a feature of tree. Among all graphs with given vertices, tree is a connected graph with the least number of edges. Tree is also graph without cycle and with the most number of edges. From this, in any graph G, if $e < n-1$, then G is disconnected; if $e > n-1$, then G must have a cycle.

Another feature of tree is also very useful, as shown below.

Example 2 If T is tree, then there is only one chain between any two vertices of T. Conversely, if there is only one chain between any two vertices of T, then T must be a tree.

Proof If T is a tree, then T is connected, and there is at least one chain between any two vertices. Since T has no cycle, there is only one chain between any two vertices of T.

Conversely, if there is only one chain between any two vertices of T, then T is clearly connected and T has no cycle. Otherwise, there would be at least two chains between any two vertices on a cycle, this introduces a contradiction.

Example 3 There are n cities, each city can call another city through some intermediate cities. Prove that there are at least $n-1$ direct lines, each of which connects two cities. (Hungary Mathematical Olympiad)

Proof Draw a graph G and denote n cities by n vertices. If there is a direct line between two cities, then join two corresponding vertices. By hypothesis, G is a connected graph. So the number of edges of G is no less than $n-1$, then there are at least $n-1$ lines connecting every two cities.

This problem can be considered in the following way. If the connected graph G obtained has a cycle, we delete an edge on the cycle and obtain a graph G_1. The number of edges of G_1 is less than that of G by one, but G_1 is still connected. If graph G_1 still has a cycle, we delete an edge on the cycle and obtain a graph G_2, and so on until the graph obtained has no cycle. The graph is certainly a tree. It has $n-1$ edges, so graph G has at least $n-1$ edges.

The above tree obtained is called the generating tree of graph G. Adding several edges to the generating tree, we can get the original graph.

Example 4 In a certain region, twenty players of a tennis club have played fourteen singles. Every person plays at least once. Prove that there are six pairs singles, in which twelve players are distinct.

Proof This question has occurred in Chapter 2. Here we prove it from the viewpoint of tree.

Denote twenty players by twenty vertices. If two persons have played a game, add an edge between them. There are fourteen edges in all, each vertex is incident to at least one edge. Our conclusion is equivalent to: it is possible to find six edges so that any two of them are not adjacent.

Suppose the graph has n connected branches, among which the ith branch has v_i vertices, e_i edges. Clearly, $e_i \geqslant v_i - 1$, so

$$\sum_{i=1}^{n} e_i \geqslant \sum_{i=1}^{n} (v_i - 1) = \sum_{i=1}^{n} v_i - n.$$

But $\sum_{i=1}^{n} e_i = 14$, $\sum_{i=1}^{n} v_i = 20$, so $14 \geqslant 20 - n$, $n \geqslant 20 - 14 = 6$. Since every vertex is incident to at least one edge, it is impossible that there exists a connected branch, which contains only one isolated vertex. Hence choose an edge from every connected branch, which promises that they are not adjacent and the number of edges is at least six. We finish the proof.

From Fig. 4.5, the number of vertices is twenty, the number of edges is fourteen. Choose arbitrarily seven edges, then there must be two edges that are in the same connected branch and adjacent. Hence six is the best possible.

Fig. 4.5

Example 5 Given $2n$ vertices on a plane, cover them by some circles. Prove that if each circle covers at least $n+1$ vertices, then any two vertices can be joined by a zigzag line on a plane, and this line is fully covered by the cycles.

Proof We denote $2n$ points by $2n$ vertices. If there is a circle, which covers two points, then adding an edge between these two vertices, we obtain a graph G. By hypothesis, the degree of each vertex is no less than n. Each edge in the graph can be fully covered by a circle. So we only need to show that G is connected.

If G is disconnected, then there exists a connected branch G_1, which has at most n vertices. We have $d(v) \leqslant n-1$ for every vertex v of G_1. A contradiction. So G is connected.

Example 6 n ($n > 3$) table tennis players play several single games. The opponents of any two players are different. Prove that we can remove one player so that the opponents of any two players in the remaining players are still different. (China Mathematical Competition in 1987)

Proof Denote n players by n vertices v_1, v_2, ..., v_n. If the proposition is not true, i.e. any player cannot be removed. For player $v_k (1 \leqslant k \leqslant n)$, since he cannot be removed, after removing v_k, we can always find a pair of players v_i and v_j, whose opponents are the same (if there are several pairs, choose any one pair). This indicates that the opponents of v_i and v_j are only different due to v_k. Without loss of generality, we suppose that v_i has played a game with v_k, but v_j has not played a game with v_k. Add an edge between v_i and v_j, label k. We obtain a graph with n vertices and n edges, and there are n edges with n different numbers.

The graph with n vertices and n edges must have a cycle. Suppose

$v_{i_1}, v_{i_2}, \ldots, v_{i_k}$ is a cycle. Along the cycle, going through an edge means adding or subtracting a player from the players and the added player and the subtracted player are different. Going through a cycle, we can return to v_{i_1}, i.e. after adding or subtracting players from the players who have played with v_{i_1}, the result is the same as the original opponents of v_{i_1}. A contradiction.

Hence, there is at least one player who can be removed in n players.

Example 7 In a lecture, there are five mathematicians. Each of them took a nap twice and every two of them took naps at the same time. Prove that there must be three persons who took naps at the same time. (USA MO 1986)

Proof Denote ten naps of five mathematicians by vertices v_1, v_2, ..., v_{10}, add an edge between v_i and v_j if and only if ith and jth took naps at a common time. We obtain a graph.

By hypothesis, every two mathematicians took naps at a common time, so graph G has at least $\binom{5}{2} = 10$ edges. But graph G has ten vertices, so G must have a cycle.

Let the cycle be $v_{i_1} v_{i_2} \ldots v_{i_k} v_{i_1}$, suppose v_{i_1} waked up first. Then as soon as v_{i_1} waked up, where v_{i_2} and v_{i_k} were still taking a nap. This proves that there must be three persons who took naps at a common time.

Example 8 There are 1990 residents in a district. Every day each of them tells the news they heard yesterday to all his acquaintances and any news can gradually be known to all. Prove that we can select 180 residents so that they are informed of some news at the same time and after at most ten days, this news will be known by all residents.

Proof Denote these residents by vertices, two adjacent vertices means that the corresponding residents are familiar. We obtain a graph G with 1990 vertices.

By hypothesis, G is connected. Let this graph be a tree T_{1990}. Otherwise replace it by its generating tree. In the tree T_{1990}, choose the longest chain, and denote it by

$$v_1^{(1)} v_2^{(1)} v_3^{(1)} \ldots v_{11}^{(1)} \ldots v_n^{(1)}.$$

Choose $v_{11}^{(1)}$ as a representative of the residents, and delete edge $(v_{11}^{(1)}, v_{12}^{(1)})$. Then T_{1990} is divided into two trees. In the first tree, the distance of every vertex v and $v_{11}^{(1)}$ is no more than 10. Otherwise, in tree T_{1990}, the chain from v to $v_p^{(1)}$ is longer than the chain from $v_1^{(1)}$ to $v_n^{(1)}$. Hence the news known to the representative $v_{11}^{(1)}$ will be known to the persons in the first tree within ten days. The latter tree also have a longest chain, note that

$$v_1^{(2)} v_2^{(2)} v_3^{(2)} \ldots v_{11}^{(2)} \ldots v_m^{(2)},$$

where $m \leqslant 1990 - 11 = 1979$. Similarly, choose $v_{11}^{(2)}$ as a representative of the residents, and delete edge $(v_{11}^{(2)}, v_{12}^{(2)})$. The tree is divided into two trees again.

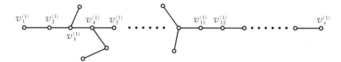

Fig. 4.6

Continue in this way. After $v_{11}^{(i)}$ ($i \leqslant 179$) has been chosen, the remaining tree has at most 11 vertices. At the same time, the number of representatives is $i + 1 \leqslant 180$. The proposition holds. Otherwise we can find the representatives

$$v_{11}^{(1)}, v_{11}^{(2)}, \ldots, v_{11}^{(179)}.$$

Each representative can tell a news to the residents in the district in ten days.

At the end, there is only one tree, which has at most

$$1990 - 11 \times 179 = 21$$

vertices. Let $v_1 v_2 \ldots v_k$ be its longest chain. If $k \geqslant 11$, then choose v_{11} as the 180th representative $v_{11}^{(180)}$. If $k < 11$, then choose v_1 as the 180th representative $v_{11}^{(180)}$. In this way, the 180 representatives are obtained

$$v_{11}^{(1)}, v_{11}^{(2)}, \ldots, v_{11}^{(179)}, v_{11}^{(180)},$$

as required.

Exercise 4

1 If the number of vertices of a connected graph G is no less than 2, then there exists at least two vertices in graph G. After removing the two vertices and their adjacent edges, the graph is still connected. (A graph without vertices is also considered as a connected graph)

2 On a coordinate plane, eleven vertical lines and eleven horizontal lines constitute a graph. The vertices of the graph are the points of intersection of the vertical and horizontal lines (lattice points), the edges are the vertical and horizontal segments between two lattice points. How many edges at least should be removed so that the degree of each vertex is less than four? How many edges at most should be deleted so that the graph keeps connected?

3 If graph G has n vertices and $n-1$ edges, then graph G is a tree. Is this proposition correct? Why?

4 A tree T has three vertices of degree 3, one vertex of degree 2 and other vertices are all pendant vertices. (1) How many pendant vertices are there in T? (2) Draw two trees which satisfy the above requirement of degrees.

5 A tree has n_i vertices whose degrees are i, $i=1, 2, \ldots, k$. If the numbers $n_2, \ldots, n_k$ are all known, what is n_1? If n_r ($3 \leqslant r \leqslant k$) is not known, and n_j ($j \neq r$) is known, what is n_r?

6 Let $d_1, d_2, \ldots, d_n$ be n positive integers, $n \geqslant 2$, and $\sum_{i=1}^{n} d_i = 2n-2$. Prove that there exists a tree where the degrees of its vertices are $d_1, d_2, \ldots, d_n$.

7 There are n ($n \geqslant 3$) segments on the plane where any three of them have a common end, then these n segments have a common end.

8 Given a number table with n rows and n columns, every two rows are different. Prove that there must exist a column, after deleting it, every two rows are still different.

9 Let V be the set of all the vertices of G, E the set of all the edges of G. Prove that if $|E| \geq |V|+4$, then there must be two cycles which have no common edges. (Pösá Theorem)

10 In an evening, 21 persons make phone calls, someone finds that these 21 persons call 102 times and every two persons call at most once. He also finds there exist m persons, the first one calls the second, the second calls the third, ..., the $(m-1)$th calls the mth, and the mth calls the first. He does not tell the value of m, only says m is odd. Prove that there exist 3 persons among 21 persons, each of the three persons calls each other.

11 A country has a number of cities. There are roads connecting the cities and each city has 3 roads connected to it. Prove that there exist roads forming a cycle whose length is not divisible by 3.

Chapter 5 Euler's Problem

Euler's problem arised from the famous seven-bridge problem. Königsberg is located in Europe and Pregel River with a beautiful view runs across the city. In the river, there are two islands A and D. On the river, there are seven bridges joining two islands and the riversides B and C (Fig. 5.1).

Fig. 5. 1

Question: Can a traveler go through every bridge once and once only? This is the famous Königsberg seven-bridge problem.

Clever Euler, using his insight, found that it was an interesting geometry problem. In 1936, he published a thesis named "seven bridges of Königsberg" and solved the seven-bridge problem. It was generally considered as the first thesis in graph theory. Euler changed Fig. 5. 1 into a graph G as shown in Fig. 5. 2. Islands A, D and riversides B, C are four vertices and seven edges in graph G denote seven bridges. Hence the seven-bridge game becomes the following problem: Can the graph be drawn without lifting one's pen and traversing every edge only once? Drawing a graph with a continuous line does not necessarily have to come back to the original

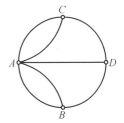

Fig. 5. 2

starting point, that is to say, the problem of drawing a graph with a continuous line (or in the end, coming back to the original starting point) is equivalent to the problem whether the graph is a chain (or a cycle).

If G is a chain from v_1 to v_{n+1}, then each vertex v_i ($i = 2, 3, \ldots, n$) different from v_1 and v_{n+1} are even vertices. Because to vertex v_i, if there is an edge entering v_i, then there is also an edge leaving v_i. The entering and leaving edges cannot be repeated. Hence there are always double edges adjacent to v_i. Therefore, graph G has at most two vertices with odd degrees, i.e. v_1 and v_{n+1}. If G is a cycle, by the above reasoning, v_1 and v_{n+1} are also even vertices. Hence, if G is a chain(cycle), then the number of vertices with odd degrees in G is equal to 2(0). This is the necessary condition that G is a chain(cycle). In other words, if the number of vertices with odd degrees in G is more than 2, then G is not a chain and it cannot be drawn without lifting one's pen.

In Fig. 5.2, A, B, C, D are all odd vertices, hence this graph is not a chain and it cannot be drawn without lifting one's pen. That is to say, it is impossible for a traveler to go through each bridge once and only once.

The following is "the solution to Euler's problem".

Theorem 1 A finite graph G is a chain or a cycle if and only if G is connected and the number of vertices with odd degree in G is equal to 2 or 0. When and only when the number of vertices with odd degrees in G is equal to 0, the connected graph G is a cycle.

Proof The necessity has been proved above, and next we prove the sufficiency.

First, we prove: If G is connected and the number of odd vertices is 0, then G must be a cycle.

Start from any vertex v_0 of G and go to v_1 through the adjacent edge e_1. Since $d(v_1)$ is even, go to v_2 through the adjacent edge e_2 from v_1, and continue in this way. Each edge is chosen only once and we must return to v_1 after several steps, so we obtain a cycle μ_1:

$v_0 v_1 \ldots v_0$.

If μ_1 is G itself, then the proposition holds. Otherwise we can find a subgraph G_1 by deleting μ_1 in G, then every vertex in G_1 is also an odd vertex. Since G is connected, then there must exist a common vertex u in μ_1 and in G_1 and a cycle μ_2 from u to u in G_1. So μ_1 and μ_2 still constitute a cycle. Repeat the above process. Since G has only a finite number of edges in all, the cycle finally obtained is graph G itself.

Now we prove the second case. Suppose G is connected, and the number of odd vertices is 2. Let u, v be the two odd vertices, add an edge e between u and v, we get graph G'. Hence the number of odd vertices is 0 in G', so G' is a cycle. Therefore, after deleting e, G is a chain.

We call a cycle in a graph an Euler tour if it traverses every edge of the graph exactly once. A graph is Eulerian if it admits an Euler tour.

Furthermore, there is the following question: If the number of odd vertices in a connected graph G is not 0 or 2, then by lifting one's pen how many times can G be drawn? We know that the number of odd vertices is even, so we have the following conclusion.

Theorem 2 If G is connected and has $2k$ odd vertices, then graph G can be drawn by lifting one's pen k times and at least k times.

Proof Divide these $2k$ odd vertices into k pairs: v_1, v'_1; v_2, v'_2; $\ldots$; v_k, v'_k, add an edge e_i between v_i and v'_i and obtain G'. Graph G' has no odd vertex, so G' is a cycle. Delete these k added edges, then this cycle is divided into at most k parts, i.e. k chains. This indicates that G can be drawn by lifting one's pen k times.

Suppose G is divided into h chains, each chain has at most two odd vertices. Hence $2h \geq 2k$, i.e. $h \geq k$. Graph G can be drawn by lifting one's pen at least k times.

Example 1 Fig. 5.3 is a plane graph of a building and there is a living room. After entering the front door into the living room, there are four other rooms. If you enter from the front door, can you enter all the rooms (including the living room) through all the doors and go through each door only once?

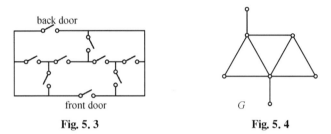

Fig. 5.3   Fig. 5.4

Proof   The answer is no.

Consider five rooms, the area outside the front door and that outside the back door can be denoted by vertices, add an edge between the corresponding vertices if these places are connected by doors and we get a graph G (see Fig. 5.4). In graph G, the number of odd vertices is 4, so G is not a chain. So the answer is no.

Example 2   Fig. 5.5 can be drawn by lifting one's pen 5 times and Fig. 5.6 can be drawn by lifting one's pen 4 times.

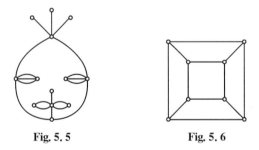

Fig. 5.5   Fig. 5.6

Example 3   On an 8×8 chessboard with 64 black and white squares we move a knight. In whatever direction the knight moves, is it possible to make the knight move to all the squares and each square only once? (A knight moves from a corner square of a rectangle with 2×3 black and white squares to another corner square diagonal to the starting position.)

Proof   Fig. 5.8 gives us an answer of the question.

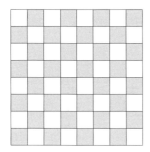

56	41	58	35	50	39	60	33
47	44	55	40	59	34	51	38
42	57	46	49	36	53	32	61
45	48	43	54	31	62	37	52
20	5	30	63	22	11	16	13
29	64	21	4	17	14	25	10
6	19	2	27	8	23	12	15
1	28	7	18	3	26	9	24

Fig. 5. 7   Fig. 5. 8

Solving this question, we often try the following four methods.

1. Each time we place the knight in a position where the number of squares where the knight can move to (though it has not moved) is the least, i. e. first move to the squares where the knight has fewer places to move to, last move to the squares where the knight has move places to move to.

2. Divide the chessboard into several parts and find a Hamiltonian chain in each part (see Chapter 6) and join them together.

3. Find several cycles on the chessboard and join them.

4. Add edges to a smaller chessboard and get a Hamiltonian chain of the whole chessboard.

Example 4 In Fig. 5. 9, two ants lie in positions A, B. Ant A tells ant B: "Let us compete and see who can first creep nine edges of this graph and arrive at E first." Ant B agrees. Suppose the two ants have the same speed and start at the same time. Which ant will arrive at E first?

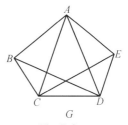

Fig. 5. 9

Proof Consider A, B, C, D, E as vertices, the original nine edges as nine edges, and we obtain graph G. Then G is connected and the number of odd vertices is 2. According to Theorem 1, it is a chain.

Since B is an odd vertex, E is also an odd vertex, there exists a chain from B to E, for example

$$BCDACEABDE.$$

For ant B, it can start from B and arrive at E along this chain.

But vertex A is an even vertex, it is impossible to go through all edges of G from A to E without repeating. Ant A must repeat at least an edge. So ant B may choose a suitable route to reach E earlier than ant A.

Example 5 As shown in Fig. 5.10, the three vertices of the big triangle are dyed in colors A, B, C. Choose some points in the interior of the big triangle, divide the big triangle into several small triangles. Each of the two small triangles has either a common vertex, or a common edge, or no common vertex at all. Color the vertices of each small triangle by one of A, B, C. Prove that whatever way you color, there must be a small triangle whose three vertices are all different.

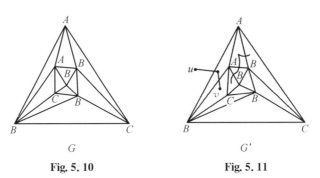

Fig. 5.10 Fig. 5.11

Proof Choose a point outside the big triangle and a point in the interior of small triangle as vertices, respectively. When two faces have a common edge AB, add an edge between the corresponding two vertices, and obtain graph G', as shown in Fig. 5.11.

A small triangle with its vertices colored A, B, C corresponds to the any vertex G' with degree 1. Other small triangles correspond to the vertices of G' with degree 0 or 2. Since the degree of a vertex outside the big triangle is 1 and the number of odd vertices is even, Therefore G' has at least an odd vertex v except u. That is to say, in Fig. 5.11, there is at least a small triangle whose three vertices have three colors A, B, C, respectively.

Note By the conclusion of this example, the famous Brouwer fixed point theorem follows.

Example 6 A graph which consists of a convex n-polygon and $n-3$ disjoint diagonal lines in the polygon is called a subdivision graph.

Prove that there exists a subdivision graph which is a cycle drawn without lifting one's pen (i.e. start from a vertex, go through each segment only once and return the starting point) if and only if $3 \mid n$. (The 5th China Mathematical Competition)

Proof First prove by induction that the condition $3 \mid n$ is sufficient.

When $n = 3$, clearly the proposition holds.

Suppose for any convex $3k$-polygon, there exists a subdivision graph which is a cycle drawn without lifting one's pen. For a convex $3(k+1) = 3k+3$-polygon $A_1 A_2 A_3 \ldots A_{3k+3}$, join $A_4 A_{3k+3}$. Since $A_4 A_5 \ldots A_{3k+3}$ is a convex $3k$-polygon, by induction, $A_4 A_5 \ldots A_{3k+3}$ contains a subdivision graph which is a cycle drawn without lifting

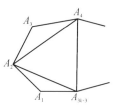

Fig. 5.12

one's pen. Construct this subdivision graph and join $A_2 A_4$, $A_2 A_{3k+3}$, so we obtain a subdivision graph of a convex $3(k+1)$-polygon $A_1 A_2 A_3 \ldots A_{3k+3}$. Since the subdivision graph of $A_4 A_5 \ldots A_{3k+3}$ is a cycle, we start from A_{3k+3}, go through each edge of the subdivision graph only once and return to A_{3k+3}. Then go through $A_{3k+3} A_1$, $A_1 A_2$, $A_2 A_3$, $A_3 A_4$, $A_4 A_2$, $A_2 A_{3k+3}$, and return to A_{3k+3}, again. This proves that for any convex $(3k+3)$-polygon, there also exists a subdivision graph, which is a cycle drawn without lifting one's pen. So the sufficiency has been proved.

Next prove the necessity. Assume that a convex n-polygon has a subdivision graph, which is a cycle drawn without lifting one's pen. Then each vertex of the graph is an even vertex. Clearly a convex quadrangle and a convex pentagon do not have a subdivision graph such that each vertex is an even vertex. Thus when $3 \leqslant n < 6$, if a convex n-polygon has a subdivision graph, such that each vertex is an even vertex, then $n = 3$. When $3 \leqslant n < 3k$ ($k > 2$), if a convex n-polygon has a subdivision graph, such that each vertex is an even vertex, then $3|n$. Now consider $3k \leqslant n < 3(k+1)$. Suppose that a convex n-polygon $A_1A_2 \ldots A_n$ has a subdivision graph, such that each vertex is an even vertex. It is easy to see that any subdivision of a convex n ($n > 3$)-polygon can divide the convex n-polygon into $n - 2$ small triangles, which have no common interior, and at least two of these triangles contain two adjacent edges of the convex n-polygon as two edges. Hence without loss of generality, let A_1A_3 be a diagonal line of a subdivision graph of the convex n-polygon $A_1A_2 \ldots A_n$ (as shown in Fig. 5.13). So A_1A_3 is still an edge of another $\triangle A_1A_3A_i$ in the subdivision graph. By hypothesis that $A_1A_2 \ldots A_n$ has a subdivision graph such that each vertex is an even vertex, hence $i \neq 4$. Otherwise, A_3 is an odd vertex. Equally $i \neq n$; otherwise, A_1 is an odd vertex. Therefore $4 < i < n$. The subdivision graph of $A_1A_2 \ldots A_n$ gives rise to subdivision graphs of a convex $(i-2)$-polygon $A_3A_4 \ldots A_i$ and a convex $(n-i+2)$-polygon $A_1A_2 \ldots A_n$, respectively. Each vertex of the convex polygons corresponding to these two subdivision graphs is even. Hence by induction,

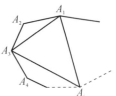

Fig. 5.13

$$3 \mid i-2, \ 3 \mid n-i+2,$$

so $3|n$. Hence the necessity has been proved.

The necessity also can be proved by the coloring method. For a subdivision graph of a convex n-polygon, we can color the divided triangles using two colors, such that two triangles with a common edge have different colors. Do as follows: draw diagonal lines in sequence so that each diagonal line divides the interior of polygon into two parts, in one part keep the original color, in another part change color. Finally, we draw all the diagonal lines and obtain the needed color.

Since convex polygon has a subdivision graph, which is a cycle drawn without lifting one's pen, each vertex is an even vertex. So the number of triangles at each vertex is odd. In the above coloring method, all the edges of the polygon belong to the triangles with the same color. Let it be black (see Fig. 5.14). Denote the number of edges of white triangles by m. Clearly $3 \mid m$, each edge of the white triangles is also that of the black triangles. However, all the edges of the polygon are those of the black triangles, so the number of edges of the black triangles is $m + n$, so $3 \mid n$.

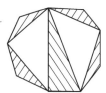

Fig. 5.14

Example 7 Suppose $n > 3$, consider the set E of $2n - 1$ distinct points on a circle. Color some points of E black, and other vertices no color. If there exists at least a pair of two black points such that between two arcs with the two black points as their endpoints we can find one of them whose interior (not including endpoints) contains exactly n points, then we call the coloring "good". If each coloring with k points of E colored black is good, find the minimum value of k. (31th International Mathematical Olympiad)

Proof Denote the points of E by $v_1, v_2, \ldots, v_{2n-1}$ according to the anti-clockwise direction and add an edge between v_i and $v_{i+(n-1)}$, $i = 1, 2, \ldots, 2n - 1$. We assume that $v_{j+(2n-1)k} = v_j$, for $k = 1, 2, 3, \ldots$. Then we get a graph G. The degree of each vertex in G is 2 (i.e. every vertex is adjacent to two other vertices) and v_i and v_{i+3} are adjacent to a common vertex. Since each vertex of G is an even vertex, G consists of one or several cycles.

(i) When $3 \mid (2n - 1)$, graph G consists of three cycles, the vertex set of each cycle is

$$\left\{ v_i \mid i = 3k, k = 1, 2, \ldots, \frac{2n-1}{3} \right\},$$

$$\left\{ v_i \mid i = 3k + 1, k = 0, 1, \ldots, \frac{2n-4}{3} \right\},$$

$$\left\{ v_i \mid i = 3k + 2, k = 0, 1, \ldots, \frac{2n-4}{3} \right\}.$$

Since the number of vertices in each cycle is $\frac{2n-1}{3}$, it is possible to choose at most $\frac{1}{2}\left(\frac{2n-1}{3} - 1\right) = \frac{n-2}{3}$ vertices and every two of them are not adjacent (note that $\frac{2n-1}{3}$ is odd). So we may choose $n-2$ vertices which are all not adjacent pairwise. By the pigeonhole principle, we must color at least $n-1$ vertices black to assure that there is at least a pair of adjacent black vertices.

(ii) When $3 \nmid (2n-1)$, each vertex of $v_1, v_2, \ldots, v_{2n-1}$ can be denoted in the form of v_{3k}. So graph G is a cycle with length $2n-1$. We can choose $n-1$ nonadjacent vertices on this cycle and at most $n-1$ nonadjacent vertices. Hence color at least $n-1$ vertices black so that there is at least a pair of adjacent black vertices.

In other words, when $3 \nmid (2n-1)$, the minimum value of k is n, and when $3 \mid (2n-1)$, the minimum value of k is $n-1$.

Exercise 5

1 What is the value n when the complete graph K_n is a cycle? What is the value n when the complete graph K_n is a chain? What are the values m, n, when the complete bipartite graph $K_{m,n}$ is a cycle?

2 Suppose graph G can be drawn by lifting one's pen at least k times, G' is obtained by deleting an edge. How many times at least can G' be drawn by lifting one's pen?

3 Determine whether each of the Fig. 5. 15 can be drawn without lifting one's pen.

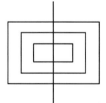 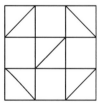

Fig. 5. 15

4 Choose arbitrarily n ($n > 2$) vertices, and join each vertex to all other vertices. Can you draw these segments without lifting one's pen, so that they join end to end and finally return to the starting point?

5 If at a conference, each person exchanges views with at least $\delta \geq 2$ persons. Prove that it is definitely possible to find k persons v_1, v_2, ..., v_k, such that v_1 changes opinion with v_2, v_2 exchanges views with v_3, ..., v_{k-1} changes opinion with v_k, and v_k exchanges views with v_1, where k is an integer greater than δ.

6 As shown in Fig. 5.16, graph G has 4 vertices, and 6 edges. They are all on a common plane. This plane is divided into 4 regions Ⅰ, Ⅱ, Ⅲ, Ⅳ, and we call them regions faces. Suppose there are two points Q_1, Q_2 on these faces. Prove that there is no line μ joining Q_1 and Q_2 which satisfies: (1) μ cuts across each edge only once; (2) μ does not go through any vertex v_j ($j = 1, 2, 3, 4$).

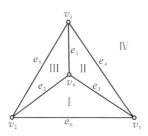

Fig. 5. 16

7 Arrange n vertices v_1, v_2, ..., v_n in order on a line. Each vertex is colored in red or blue. If the ends of a segment $v_i v_{i+1}$ are colored differently, we call it a standard segment. Suppose the colors of v_1 and v_n are different. Prove that the number of the standard segments is odd.

8 Choose some points on the edges and in the interior of $\triangle ABC$. Divide $\triangle ABC$ into various small triangles. Each two small triangles has either a common vertex, or a common edge, or no common vertex at all. Use A, B or C to label those vertices in the interior of $\triangle ABC$. Use A or B to label the vertices on the edge AB of the big triangle, label B or C to the vertices on the edge BC of the big triangle, and label C or A to the vertices on the edge CA of the big triangle. Prove that there must be a small triangle, whose three vertices are A, B, C.

9 In the following figure with 25 small squares, try to design a walk starting from point A, going through the edges of all the small squares and finally returning to A, such that the path is the shortest.

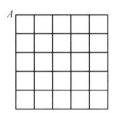

Fig. 5. 17

Chapter 6 Hamilton's Problem

In 1856, the famous British mathematician Willian Rewan Hamilton brought forward a game whose name was "go around the world". He denoted twenty big cities by twenty vertices of a regular dodecahedron. You should go along the edges, pass through every city once and at last return to the starting point.

The game was welcomed all around the world. In this game, we see a chain that it passes through every vertex only once. We call this chain (cycle) a *Hamiltonian chain* (cycle). If a graph contains a Hamiltonian cycle, we call it a *Hamiltonian graph*.

On the surface, Hamilton's problem is similar to Euler's problem. But, in fact, they are different in nature. Hamilton's problem is one difficult problem in graph theory that has not been solved. Until now we can not find a necessary and sufficient condition to characterize it. So we have different methods for different problems. We shall use some examples to illustrate.

Example 1   Does Fig.6.1 contain a Hamiltonian chain or a Hamiltonian cycle?

Solution   As Fig.6.1 shows us, according to numbers shown we can find a Hamiltonian cycle.

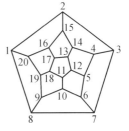

Fig. 6.1

Here we use the "direct search" method to solve the problem of "go around the world". That is, go from a vertex and search one by one in order to find the Hamiltonian chain (cycle). If we find one chain, we have found one solution. If not, there does not exist a solution.

This method can always be used on simpler graphs and often to those graphs which contain a Hamiltonian chain (cycle).

Example 2 In an international mathematics conference, there are seven mathematicians come from different countries. The language they can speak is

    A: English

    B: English and Chinese

    C: English, Italian and Spanish

    D: Chinese and Japanese

    E: German and Italian

    F: French, Japanese and Spanish

    G: French and German

How can we arrange these seven mathematicians round a table so that everyone can talk with the person beside him?

Solution We denote the seven mathematicians by seven vertices A, B, C, D, E, F, G. If two persons can speak a common language, then we join the vertices representing them and we get a graph G. As Fig. 6.2 shows us, the problem of arranging seats becomes a problem of finding a Hamiltonian cycle. Arrange the seats in the order of the cycle, so that everyone can talk with the person beside him.

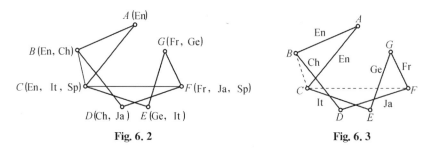

Fig. 6. 2   Fig. 6. 3

Note Ch = Chinese, En = English, Fr = French, Ge = German, It = Italian, Ja = Japanese, Sp = Spanish.

In Fig. 6.2, we draw a cycle in a bold line and then we get our solution, which also means if we arrange the seats in the order A, B, D, F, G, E, C, everyone can talk to the persons beside him. The common language is labelled on each corresponding edge in Fig. 6.3.

Example 3 Determine whether the graph G in Fig. 6.4 contains a Hamiltonian chain or cycle?

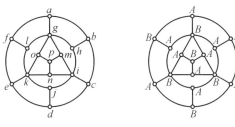

Fig. 6.4　　　　Fig. 6.5

Solution We mark one vertex in graph G as A. For example, we mark the vertex a as A and all the vertices adjacent to the vertex a as B all the vertices adjacent to B as A. Then we mark the vertices adjacent to the vertex which is marked B as A and the vertices adjacent to the vertex which is marked A as B until we mark all the vertices. As Fig. 6.5 shows us, if G contains a Hamiltonian cycle, the cycle must go through A and B in turn. So the difference between the numbers of A and B is no more than 1. But in Fig. 6.5, there are nine A vertices and seven B vertices. The difference is 2, so there are no Hamiltonian chain.

Generally, to a bigraph $G = (V_1, V_2, E)$, there is a simple method to see whether the graph contains a Hamiltonian chain or a Hamiltonian cycle.

Theorem 1 In a bigraph $G = (V_1, V_2, E)$, if $|V_1| \neq |V_2|$, G must contain no Hamiltonian cycle. If the difference between $|V_1|$ and $|V_2|$ is more than 1, G must contain no Hamiltonian chain.

We can use the same method as Example 3 to prove it.

Example 4 Fig. 6.6 shows us half of a chessboard. A knight is at the bottom right corner. Can the knight move along every square continually once only? What happens if we delete the black panes at

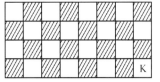

Fig. 6.6

the two corners of the half chessboard?

Solution We consider the following graph. We denote the squares in the half of a chessboard by the vertices of a graph. If a knight can move along from one square to another square in one step, we join the two vertices representing the panes. So the problem becomes a new problem of determining whether the graph contains a Hamiltonian chain. In the graph, whether the two vertices are adjacent is determined by the rule of how a knight moves. Two vertices are adjacent if they are at the two ends of the shape of letter "L" on the chessboard. Color a vertex by the color of the square representing the vertex in the chessboard, the colors of the two adjacent vertices are always different. So between the two adjacent vertices in the graph, one vertex is black and the other is white. The number of the black vertices is the same as the number of white vertices, so there exists a Hamiltonian chain. We can use the trail and error to find a chain.

15	18	7	22	11	28	5	24
8	21	16	27	6	23	2	29
17	14	19	10	31	12	25	4
20	9	32	13	26	3	30	1

Fig. 6.7

Now let us consider the second part of the problem. Again we use the above method to convert the problem to determining whether the graph contains a Hamiltonian chain. The number of the black squares is 14 and the number of white squares is 16. According to Theorem 1, the graph contains no Hamiltonian chain. It means that the knight cannot move along every square continually once only when we delete the black squares.

Now we do not know the necessary and sufficient condition of the problem, namely, whether a connected graph contains a Hamiltonian chain (cycle). However many first-class mathematicians have done some hard work for more than one century, they have found some sufficient conditions and some necessary conditions. In what follows we give a sufficient condition for the problem whether a simple graph contains a Hamiltonian chain.

Theorem 2 G is a simple graph with n ($n \geq 3$) vertices. For every pair of vertices v, v',

$$d(v) + d(v') \geq n - 1,$$

then G contains a Hamiltonian chain.

Proof First, we prove that G is a connected graph. Suppose that G contains two or more connected components. Suppose one of them has n_1 vertices, and another has n_2 vertices. We take one vertex each, v_1 and v_2, from the two components. Then $d(v_1) \leq n_1 - 1$ and $d(v_2) \leq n_2 - 1$. So

$$d(v_1) + d(v_2) \leq n_1 - n_2 - 2 < n - 1.$$

It contradicts the hypothesis. So G is a connected graph.

Now we prove that there exists a Hamiltonian chain. The method of proof in fact gives us a method to construct a Hamiltonian cycle.

Suppose G contains a chain from v_1 to v_p: $v_1 v_2 \ldots v_p$. If v_1 or v_p is adjacent to one vertex which does not lie in the chain, we can extend this chain so that the vertex lies in the chain. Otherwise, v_1 and v_p are only adjacent to the vertices of the chain. There must exist a cycle with vertices $v_1, v_2, \ldots, v_p$. Suppose the vertex set adjacent to v_1 is $\{v_{j1}, v_{j2}, \ldots, v_{j_k}\}$. Here $v_{j1}, v_{j2}, \ldots, v_{j_k}$ are the vertices on the chain and $p < n$.

If v_1 is adjacent to v_p, then there exists a cycle $v_1 v_2 \ldots v_p v_1$.

If v_1 is not adjacent to v_p, then there exists a vertex $v_l (2 \leq l \leq p)$ which is adjacent to v_1 and v_{l-1} is adjacent to v_p as Fig. 6.8 shows us. If not, v_p is adjacent to $p - k - 1$ vertices at most, which excluding $v_{j_1-1}, v_{j_2-1}, \ldots, v_{j_k-1}$ and v_p, means that

$$d(v_1) + d(v_p) \leq k + (p - k - 1) = p - 1 < n - 1.$$

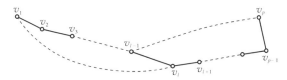

Fig. 6.8

It contradicts the hypothesis. So there exists a cycle containing v_1, $v_2, \ldots, v_p$: $v_1 v_l v_{l+1} \ldots v_p v_{l-1} v_{l-2} \ldots v_2 v_1$.

If $p = n$, already there exists a Hamiltonian cycle. If $p < n$, since G is connected, G must contain a vertex v' which does not belong to the chain and is adjacent to one vertex v_k of $v_1 v_2 \ldots v_p$. As Fig. 6.9 shows us, we can get a cycle containing $v_1, v_2, \ldots, v_p, v'$: $v' v_k v_{k+1} \ldots v_{l-1} v_p v_{p-1} \ldots v_l v_1 v_2 \ldots v_k v'$. Repeat above process until there exists a chain with $n-1$ edges.

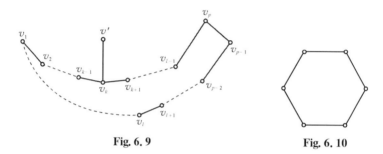

Fig. 6.9   Fig. 6.10

It is easy to see that the condition of Theorem 2 about the existence of a Hamiltonian chain is sufficient, but not necessary. Suppose G is a polygon of n sides, as Fig. 6.10 shows us. Here $n = 6$. Though the sum of the degrees of any two vertices is $4 < 6 - 1$, G contains a Hamiltonian chain.

In 1960, Ore gave us a sufficient condition for a Hamiltonian graph in *American Mathematical Monthly*.

Theorem 3 G is a simple graph with n vertices. For every two nonadjacent vertices v, v'

$$d(v) + d(v') \geq n,$$

then G is Hamiltonian cycle.

Proof When $n = 3$, by the given condition we know that G must be a complete graph K_3. The conclusion is true.

Suppose $n \geq 4$, we prove by contradiction. Suppose G with n vertices satisfies the given condition of degree and contains no Hamiltonian cycle.

Without loss of generality, G is the graph which satisfies the given condition and contains the most number of edges. In other words, after adding an edge to G, G contains a Hamiltonian cycle. Otherwise, G can be added several edges until we can not add edges. After adding edges, the degrees of vertices satisfy the condition of degree. Then we get a Hamiltonian chain contains every vertex of G. We denote the chain by $v_1 v_2 \ldots v_n$, then v_1 is not adjacent to v_n. So

$$d(v_1) + d(v_n) \geq n.$$

Then among $v_2, v_3, \ldots, v_{n-1}$, there must be a vertex v_i so that v_1 is adjacent to v_i and v_n is adjacent to v_{i-1} as Fig. 6.11 shows us. Otherwise, there are $d(v_1) = k$ vertices $v_{i_1}, v_{i_2}, \ldots, v_{i_k}$ ($2 \leq i_1 \leq i_2 \leq \cdots \leq i_k \leq n-1$) adjacent to v_1, and v_n is not adjacent to v_{i_1-1}, $v_{i_2-1}, \ldots, v_{i_k-1}$. So

$$d(v_n) \leq n - 1 - k,$$

then

$$d(v_1) + d(v_n) \leq k + n - 1 - k = n - 1 < n,$$

which contradicts the condition. So G contains a Hamiltonian cycle $v_1 v_2 \ldots v_{i-1} v_n v_{n-1} \ldots v_i v_1$, which also contradicts the hypothesis. We complete the proof.

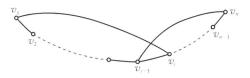

Fig. 6.11

For the complete graph K_n ($n \geq 3$), clearly there is Hamiltonian cycle.

Example 5 n persons take part in a conference. During the conference time, everyday they must sit at a round table to have dinner. Every evening, every person must sit beside different persons. How many times at most will there be such dinners?

Solution We denote n persons by n vertices. Draw a complete graph K_n, then the Hamiltonian cycle in K_n is a way of sitting round the table. The most number of times is equal to the number of Hamiltonian cycle with no common edges in K_n.

K_n contains $\frac{1}{2}n(n-1)$ edges and every Hamiltonian cycle contains n edges. There are at most $\left[\frac{n-1}{2}\right]$ Hamiltonian cycle with no common edges. When $n = 2k + 1$, we arrange the vertices 0, 1, 2, ..., $2k$ as Fig. 6.12. First, we take a Hamiltonian cycle (0, 1, 2, $2k$, 3, $2k-1$, 4, ..., $k+3$, k, $k+2$, $k+1$, 0), then rotate $\frac{\pi}{k}$, $\frac{2\pi}{k}$, ..., $(k-1)\frac{\pi}{k}$ clockwise around the 0 and get $k = \left[\frac{n-1}{2}\right]$ Hamiltonian cycle with no common edges. If $n = 2k + 2$, add a vertex v in the center and also get k Hamiltonian cycle.

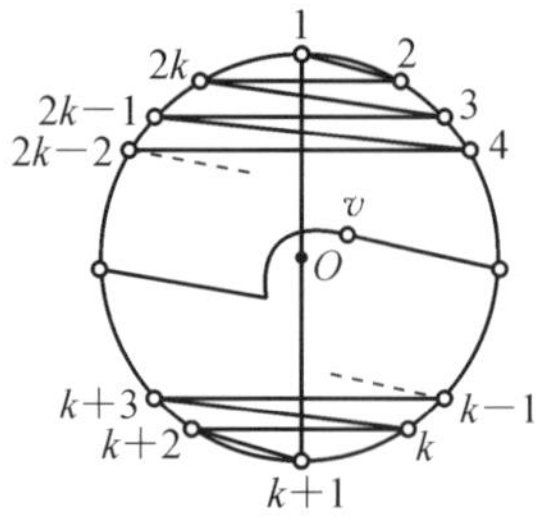

Fig. 6.12

From Theorem 3 we can induce next Theorem 4, which is given by the mathematician Dirac in 1952.

Theorem 4 G is a simple graph with n vertices. If the degree of every vertex v is no less than $\frac{n}{2}$, then G must contain a Hamiltonian cycle.

Example 6 Arrange 7 examinations in 7 days so that the two courses which are taught by one teacher cannot be arranged in two consecutive days. If every teacher can teach at most 4 courses which have examination. Prove that it is possible to arrange the examinations in the above way.

Solution Suppose the graph G contains 7 vertices. Every vertex represents an examination. If two courses are not taught by one teacher, then we join the two vertices representing the two courses. Since the number of courses taught by one teacher is no more than 4, the degree of every vertex is at least 3. The sum of degrees of any two vertices is at least 6. By Theorem 2, G contains a Hamiltonian chain. Since in this chain every two vertices which are adjacent represent two courses which are not taught by one teacher, we can arrange 7 examinations in the order of vertices in this chain.

Example 7 A factory produces two-color cloth using 6 distinct colored yarns. Among six kinds of yarns, every color must be matched in groups with three other colors. Prove that it is possible to choose three kinds of two-color cloth so that all 6 colors are present.

Proof We denote six colored yarns by six vertices. If two colors are in the same group, then we join the vertices of the two colors, and we get a graph G. What we know is that every color can be matched in groups with three other colors. For any vertex v_i, $d(v_i) \geqslant 3$, what we will prove is that graph G contains three edges, any two edges of which contain no common end.

For any vertex v in the graph G, by Theorem 4, G contains a Hamiltonian cycle which we denote by $v_1 v_2 v_3 v_4 v_5 v_6 v_1$. Then edges (v_1, v_2), (v_3, v_4), (v_5, v_6) are the three edges which contain no common vertex.

We always use the fact that the graph satisfies the sufficient condition to determine that the graph is a Hamiltonian graph and the fact that the graph does not satisfy the necessary condition to determine that the graph is not a Hamiltonian graph. Next we give a necessary condition.

Theorem 5 If G contains a Hamiltonian graph, remove several vertices $v_1, v_2, \ldots, v_k$ and their adjacent edges from the G to get a new graph G', then the number of connected components in G' is no more than k.

Proof Suppose c is a Hamiltonian cycle in G. After removing several vertices $v_1, v_2, \ldots, v_k$ and their adjacent edges from G, the cycle c can be divided into at most k parts. So the number of connected components in G' is at most k.

Example 8 Prove that there is no Hamiltonian graph in

Fig. 6. 13.

Proof Remove the vertices v_1, v_2 and their adjacent edges from Fig. 6. 13. We obtain G' with three connected components. It does not satisfy the necessary condition of Theorem 5. So there is no Hamiltonian graph in Fig. 6. 13.

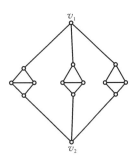

Fig. 6. 13

Finally, we give another example to end this chapter.

Example 9 If $A_0 A_1 A_2 \ldots A_{2n-1}$ is a regular polygon of $2n$ sides. Join all its diagonals to get a graph G. Prove that every Hamiltonian cycle of G must contain two edges which are paralleled in the graph.

Proof Suppose $A_i A_j$ is parallels to $A_k A_l$. Since the number of the vertices between A_i and A_l is equal to the number of the vertices between A_j and A_k, $i - l = k - j$. The sufficient and necessary condition for $A_i A_j$ parallel to $A_k A_l$ is:

$$i + j \equiv k + l \pmod{2n}.$$

Suppose $A_{i_0} A_{i_1} \ldots A_{i_{2n-1}}$ is a Hamiltonian cycle. $i_0, i_1, \ldots, i_{2n-1}$ is a rearrangement of $0, 1, \ldots, 2n$. Among them any two edges are not paralleled. So among the $2n$ numbers $i_0 + i_1, i_1 + i_2, i_2 + i_3, \ldots, i_{2n-1} + i_0$ any two numbers are not congruent module $2n$. That is, the above $2n$ numbers is in a surplus system of module $2n$. Then

$$(i_0 + i_1) + (i_1 + i_2) + (i_2 + i_3) + \cdots + (i_{2n-1} + i_0)$$
$$= 0 + 1 + 2 + \cdots + 2n - 1 = 2n^2 - n$$
$$\equiv n \pmod{2n}.$$

On the other hand,

$$(i_0 + i_1) + (i_1 + i_2) + (i_2 + i_3) + \cdots + (i_{2n-1} + i_0)$$
$$= 2(i_0 + i_1 + i_2 + \cdots + i_{2n-1})$$
$$= 2(0 + 1 + 2 + \cdots + 2n - 1) = 2n^2 - n$$
$$= 2n(2n - 1)$$
$$\equiv 0 \pmod{2n}.$$

We get two results which contradict each other. So the proof is complete.

Exercise 6

1 What is the value n so that the complete graph K_n is a Hamiltonian graph? What are the values m, n so that the complete bigraph $K_{m,n}$ is a Hamiltonian graph?

2 The graph representing the regular tetrahedron, hexahedron, octahedron or icosahedron is a Hamiltonian graph.

3 We use paper to construct an octahedron. Can we cut it into two parts so that every face is also cut into two parts and the cutting lines do not go through the vertices of the octahedron?

4 A mouse eats the cheese whose size is the same as $3 \times 3 \times 3$ cube. The way to eat it is to get through all the 27 of the $1 \times 1 \times 1$ subcube. If the mouse begins from one corner, then goes to the next subcube which has not been eaten. Can the mouse be at the center when he has eaten the cheese.

5 We divide 6 persons into 3 groups to finish 3 missions. There are 2 persons in every group. Everyone can cooperate with at least 3 persons among the other 5 persons. (1) Can the two persons of every group cooperate with each other? (2) How many distinct grouping 6 persons into 3 groups can you give?

6 A king has $2n$ ministers, among whom there are several ministers hate each other. But the number of persons every minister hate is no more than $n-1$. Can they sit in a round table so that no two adjacent ministers hate each other?

7 Among 9 children, every child knows at least four children. Can these children be arranged in a line so that every child know the child beside him?

8 A chef uses eight materials to do the cooking. He should use two materials for each dish. Every material should be used in at least four dishes. Can this chef cook four dishes so that he uses 8 distinct materials?

9 All the subsets of a finite set can be arranged side by side in a way so that every two adjacent subsets differ only in one element.

10 In a plane, a graph with n vertices and several edges is not a Hamiltonian graph. But if we remove one vertex and its adjacent edges from the graph, the graph will become a Hamiltonian graph. Find the minimum n.

11 Around a table there sit at least five persons and it is possible to rearrange their seats so that beside everyone there are two new neighbours.

Chapter 7 Planar Graph

A graph is called a *planar* graph, if it can be drawn in the plane so that its edges intersect only at their ends.

Some graphs seem to have edges intersecting, but it is not clear that they are not planar graphs. See Fig. 7.1(1), it is isomorphic to Fig. 7.1(2). So it is easy to see that Fig. 7.1(1) is a planar graph.

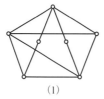 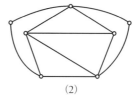

(1)   (2)

Fig. 7.1

When we talk about a planar graph G, we always assume that the planar graph is drawn according to such requirement. The vertices and edges of a graph partition the plane into several separated regions. Each area is called a *face* of G. The only unbounded face is called the *outer face* and the others are the *inner faces*. In Fig. 7.2, F_1, F_2, F_3, F_4 are inner faces, F_5 is the outer face.

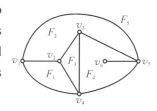

Fig. 7.2

We know Euler's Formula of a convex polyhedron in middle schools. That is, if a convex polyhedron has v vertices, e edges and f faces, then $v - e + f = 2$. We can generalize it to planar graphs.

Theorem 1 (Euler's Formula) For a connected planar graph G with v vertices, e edges and f faces, then

$$v - e + f = 2.$$

Proof We prove by induction on e.

If G has only one vertex, then $v = 1$, $e = 0$, $f = 1$, the assertion $v - e + f = 2$ holds.

If G has one edge, then $v = 2$, $e = 1$, $f = 1$, then the assertion $v - e + f = 2$ holds.

Suppose that it is true for all connected planar graphs with k edges, i.e. $v_k - e_k + f_k = 2$. Now we discuss the case when G has $k + 1$ edges.

If we add a new edge to a connected graph G with k edges such that G is still a connected graph, there must be two cases.

(i) Add a new vertex v' which is adjacent to a vertex v of G as Fig. 7.3(1) shows us. Then both v_k and e_k increase by 1, f_k does not change. So

$$(v_k + 1) - (e_k + 1) + f_k = v_k - e_k + f_k = 2.$$

(ii) Add a new edge to join two vertices of G as Fig. 7.3(2) shows us. Then both f_k, e_k are added by 1, v_k does not change. So

$$v_k - (e_k + 1) + (f_k + 1) = v_k - e_k + f_k = 2.$$

By the induction hypothesis, the theorem holds for every positive integer e.

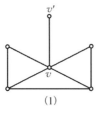
(1)
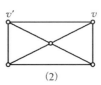
(2)

Fig. 7.3

We can use Euler's Formula to determine the maximum number of edges in a simple planar graph. Since a face has at least 3 edges, the boundaries of f faces have at least $3f$ edges. In addition, one edge belongs to the boundaries of at most 2 faces. So $2e \geqslant 3f$, and $f \leqslant \frac{2}{3}e$.

Use Euler's Formula

$$2 = v - e + f \leqslant v - e + \frac{2}{3}e,$$

that is

$$e \leqslant 3v - 6.$$

This proves the following theorem.

Theorem 2 For a connected simple planar graph G with v ($v \geqslant 3$) vertices and e edges, then

$$e \leqslant 3v - 6.$$

In fact, Theorem 2 also holds for disconnected simple planar graphs. Theorem 2 can be used to determine whether a graph is a planar graph or not.

Example 1 Prove that the complete graph K_5 is not a planar graph.

Proof Since $v = 5$, $e = 10$ do not satisfy $e \leqslant 3v - 6$. So K_5 is not a planar graph.

Example 2 Prove that $K_{3,3}$ is not a planar graph.

Proof Suppose that $K_{3,3}$ is a planar graph. Since we choose 3 vertices randomly in $K_{3,3}$, there must be 2 vertices which are not adjacent to each other. Therefore, each face has at least 4 edges as its boundary. By

$$4f \leqslant 2e, \ f \leqslant \frac{e}{2}.$$

Use Euler's Formula

$$2 = v - e + f \leqslant v - e + \frac{e}{2},$$

that is

$$e \leqslant 2v - 4.$$

In $K_{3,3}$, $v = 6$, $e = 9$, so $9 > 2 \times 6 - 4$, this leads to a contradiction. So $K_{3,3}$ is not a planar graph.

Although Euler's Formula and the inequality derived from it can be used to prove that a graph is not a planar graph, but it can do nothing to prove that a graph is a planar graph. The Polish mathematician Kuratowski showed us a brief result in 1930. Every nonplanar graph has either K_5 or $K_{3,3}$ as its subgraph. To depict the result clearly, we give the definition of homeomorphism first.

Two graphs G_1, G_2 are called *homeomophic*, if G_1 can be obtained by inserting some new vertices on G_2's edges. The two graphs in Fig. 7.4 are homeomorphic. By the definition of homeomorphism, we know that inserting or deleting a number of 2-degree vertices does not change the planarity. Kuratowski's theorem is as follows.

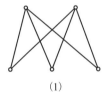

(1)　　　　　　(2)

Fig. 7.4

Theorem 3 A graph is planar if and only if it contains no subgraph homeomorphic to K_5 or $K_{3,3}$.

Although the theorem is basic, but the proof is too long to show.

Example 3 Are Fig.7.5, Fig.7.6 planar graphs?

Proof Fig. 7.5 contains a K_5 as its subgraph, and Fig. 7.6 includes a $K_{3,3}$ as its subgraph. By Kuratowski's Theorem, it is easy to see that neither of them is a planar graph.

Fig. 7.5　　　　　　Fig. 7.6

Example 4 How many kinds of regular polyhedrons are there? Can you give their numbers of edges, vetices and faces? How many edges is a vertex incident to?

Proof As each vertex in a regular polyhedron lies on at least 3 faces, when every internal angle at a vertex in a regular polyhedron is no less than 120°, it cannot be a vertex of a regular polyhedron. Therefore, we can only consider polyhedrons with regular pentagons, squares and regular triangles as faces.

(1) Polyhedron constructed from regular pentagons.

Since the internal angle of a regular pentagon is $\frac{3}{5}\pi$, and $\frac{3}{5}\pi \times 4 > 2\pi$, each vertex is 3-degree. Hence $3v = 2e$, $\frac{5}{2}f = e$, and by Euler's Formula

$$\frac{2}{3}e - e + \frac{2}{5}e = 2.$$

It is easy to see that

$$e = 30, \ v = 20, \ f = 12.$$

So, there is only one regular polyhedron constructed from regular pentagons. It is a regular dodecahedron with 20 vertices, 30 edges, and each vertex incident to 3 edges.

(2) You may prove polyhedrons constructed from squares and regular triangles by yourselves. There are 4 kinds as follows shown in Fig. 7.7. Therefore, there are 5 kinds of regular polyhedrons.

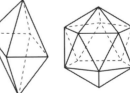

regular dodecahedron   cube   regular tetrahedron   regular octahedron   regular isosahedron

Fig. 7.7

Example 5 We partition a square into n convex polygons, where n is fixed. What is the largest number of edges that a convex polygon could have?

Proof By Euler's Formula, we know that when a convex polygon is partitioned into n polygons, then $v-e+n=1$ (because $f=n+1$).

As we partition a square into n convex polygons, for each vertex of these polygons, if it does not belong to the square, then it must be the vertex of at least 3 convex polygons. We use A, B, C, D to denote the vertices of the square, and v is an arbitrary vertex except A, B, C, D, then

$$d(v) \leqslant 3(d(v)-2).$$

Calculating the sum of all the vertices except A, B, C, D,

$$2e-(d(A)+d(B)+d(C)+d(D))$$
$$\leqslant 3(2e-(d(A)+d(B)+d(C)+d(D))-6(v-4)).$$

So,

$$4e \geqslant 2(d(A)+d(B)+d(C)+d(D))+6(v-4).$$

Since $d(A) \geqslant 2, d(B) \geqslant 2, d(C) \geqslant 2, d(D) \geqslant 2$, hence

$$2e \geqslant 8+3(v-4).$$

Using

$$v-e+n=1,$$

we can obtain the following result

$$3(e+1) = 3v+3n \leqslant 2e+4+3n,$$

that is, $e \leqslant 3n+1$.

Draw $n-1$ lines crossing an edge of a square such that all the lines are parallelled to their adjacent edge and divide the square into n rectangles. The number of edges is

$$4+3(n-1) = 3n+1.$$

In summary, the largest edge number is $3n+1$.

In 1968, two Soviet mathematicians Kozyrev and Grinberg gave a necessary condition to planar graphs with Hamiltonian cycles.

Theorem 4 If a planar graph has a Hamiltonian cycle c, let f'_i be the number of i-polygons inside c, and f''_i be the number of i-polygons outside c, then

(1) $1 \cdot f'_3 + 2 \cdot f'_4 + 3 \cdot f'_5 + \cdots = n - 2$;

(2) $1 \cdot f''_3 + 2 \cdot f''_4 + 3 \cdot f''_5 + \cdots = n - 2$;

(3) $1 \cdot (f'_3 - f''_3) + 2 \cdot (f'_4 - f''_4) + 3 \cdot (f'_5 - f''_5) + \cdots = 0$, where n is the number of the vertices of G, and also the length of c.

Proof Suppose there are d edges inside c. As G is a planar graph, its edges do not intersect and one edge divides the face into two parts. Try to think of these edges putting side by side in the graph. We get one more face each time we put an edge in the graph. So d edges divide the interior of c into $d + 1$ faces. The total number of the faces inside c is

$$f'_2 + f'_3 + f'_4 + f'_5 + \cdots = d + 1. \tag{1}$$

We can mark each i-polygon inside c with number i. The sum of all the numbers marked on the faces is the number of the edges that form the faces. Each edge inside c is counted twice, but the edges on c is counted once. So

$$2 \cdot f'_2 + 3 \cdot f'_3 + 4 \cdot f'_4 + 5 \cdot f'_5 + \cdots = 2d + n. \tag{2}$$

Subtracting equation (1) twice from equation (2), we can get

$$1 \cdot f'_3 + 2 \cdot f'_4 + 3 \cdot f'_5 + \cdots = n - 2. \tag{3}$$

Similarly, we can get the following result

$$1 \cdot f''_3 + 2 \cdot f''_4 + 3 \cdot f''_5 + \cdots = n - 2. \tag{4}$$

Equation (4) subtracted from equation (3), you can get

$$1 \cdot (f'_3 - f''_3) + 2 \cdot (f'_4 - f''_4) + 3 \cdot (f'_5 - f''_5) + \cdots = 0.$$

Example 6 Prove that the Grinberg graph (see Fig. 7.8) has no Hamiltonian cycle.

Proof Suppose that the Grinberg graph includes a Hamiltonian cycle. Since there are only pentagon, octagon and enneagon, from Theorem 4,

$$3 \cdot (f_5' - f_5'') + 6 \cdot (f_8' - f_8'') + 7 \cdot (f_9' - f_9'') = 0.$$

That is

$$7(f_9' - f_9'') \equiv 0 \pmod{3}.$$

It contradicts $f_9' - f_9'' = 1$. Therefore, the Grinberg graph has no Hamiltonian cycle.

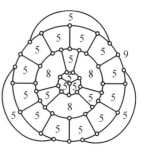

Fig. 7.8

Exercise 7

1 Let G be a simple planar graph, then it must have a vertex whose degree is no more than 5.

2 Prove that a simple planar graph with edges less then 30 must have a vertex whose degree is no more than 4.

3 Prove that in a simple planar graph with 6 vertices and 12 edges, each face is surrounded by 3 edges.

4 G is a graph with 11 or more vertices. $\overline{G}$ is the complement of G, that is, $\overline{G}$ has the same vertices as G and all the possible edges but not in G. Prove that G or $\overline{G}$ is not a planar graph.

5 Divide the plane into f parts, and every two parts are adjacent. What is the largest value of f?

6 Suppose that each vertex in a convex polyhedron is adjacent to all the other vertices. Prove that except tetrahedron such convex polyhedron does not exist.

7 In a bus network there are n stations. Each station is connected to at least 6 roads. Prove that there must be two roads intersecting on the plane.

8 If a polyhedron containing n edges exists, what is the value of n?

9 A convex polyhedron has $10n$ faces. Prove that there are n faces which have the same number of edges.

10 Prove that the graph in Fig. 7.9 has no Hamiltonian cycle.

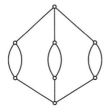

Fig. 7.9

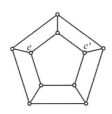

Fig. 7.10

11 The graph in Fig. 7.10 contains a Hamiltonian cycle. Prove that for any Hamiltonian cycle, if it contains edge e, then it must not contain edge e'.

12 Let $S = \{x_1, x_2, \ldots, x_n\}$ ($n \geqslant 3$) be the set of vertices on the plane. The distance between two arbitrary vertices is no less than 1. Prove that there are at most $3n - 6$ pairs of vertices whose distances are 1.

Chapter 8 Ramsey's Problem

Usually, Ramsey's problem refers to problems involving coloring, Ramsey number (Ramsey, an English logician) and the Pigeonhole Principle.

First we begin with a mathematical Olympiad problem. The problem appeared in a Hungarian mathematics competition in 1947.

Example 1 Prove that among every six persons, you can always find three persons who know each other or do not know each other.

We denote six persons by six vertices. If there are two persons who know each other, we join the corresponding vertices and color it red. If there are two persons who do not know each other, we join the corresponding vertices and color it blue. What we are to prove is that there must exist a monochromatic triangle. That is, there is a triangle whose edges are either all red or all blue.

The above is not the only one such question. A similar problem appeared in Putnam Mathematics Competition: In the space, there are six points among which any three points are not on a line and any four points are not on a plane. We join six points in pairs to get 15 line segments. We use red and green to color these line segments. (A line segment can only be colored in one color.) Prove that in whatever way the line segments are colored, there must be a monochromatic triangle.

Next let us begin our proof to the Putnam competition problem which can be regarded as a standard pattern.

Proof Let $A_1, A_2, \ldots, A_6$ be the six vertices given. Consider 5 line segments $A_1A_2, A_1A_3, \ldots, A_1A_6$ adjacent to A_1. Since there are only two colors coloring the 5 line segments, there must be 3 line segments in one color. Without loss of generality, let the three line segments be A_1A_2, A_1A_3, A_1A_4 which are colored red. (We denote red by a solid line and blue by a dotted line.) If $\triangle A_2A_3A_4$ are blue (see Fig. 8.1), then it is a monochromatic triangle. If in $\triangle A_2A_3A_4$, there are at least one edge, for example, A_2A_3 is red (as Fig. 8.2 shows us), then $\triangle A_1A_2A_3$ is a monochromatic triangle. In other words, in either case there is a monochromatic triangle.

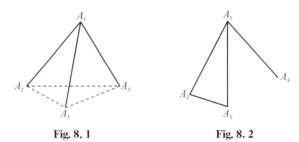

Fig. 8. 1   **Fig. 8. 2**

From this example, we can easily know that when $n \geq 6$, we use two colors to color all the edges of K_n which is called a two-color complete graph K_n for short. Then there must exist a monochromatic triangle.

Fig. 8.3 shows us a complete graph K_5 in two colors, which contains no monochromatic triangle[b].

In summary, we can get the next conclusion.

Fig. 8. 3

Theorem 1   If a complete graph K_n in two colors contains a monochromatic triangle, then the minimum n is equal to 6.

Example 2   Prove that it is impossible to color the K_{10} in four colors so that any subgraph K_4 of K_{10} contains all four colors.

Proof   We prove by contradiction. Suppose that we can find a coloring satisfying the requirement.

---

b) Vice versa, it is easy to prove that the two-color complete graph K_5 without monochromatic triangle consists of two pentagon with different colors. In other words, in the two-color complete graph K_5, if there are no blue triangle and pentagon, then there must be a red triangle.

If a vertex is adjacent to 4 edges with the same color, say blue, we color AB, AC, AD, AE all in blue. Among the edges joining B, C, D, E there must be an edge in blue. Let it be BC. Then there are 4 blue edges joining A, B, C, D, and the remaining two edges should be colored in three colors. A contradiction. So A must be adjacent to at most 3 monochromatic edges and there must be one color coloring three edges. Without loss of generality, suppose that AB, AC, AD are all blue. There are 6 edges joining A, B, C, D, so the remaining three edges are colored in different colors. There are no blue edges in BC, BD, CD.

Consider the remaining six vertices. By Theorem 1, the graph must contain a blue triangle or a triangle without blue edges.

If there are three vertices E, F, G with no blue edges joining them, then there is no blue edges joining A, E, F, G. It is a contradiction. So without loss of generality, let $\triangle EFG$ be a blue triangle. Since there are no blue edges joining B, C, D, E, there must be one blue edge among BE, CE, DE. Suppose that BE is blue, then there must exist four blue edges joining B, E, F, G. It is also a contradiction. In summary, the proposition is true.

Using the above conclusion, we can solve the following problem in the 33th International Mathematical Olympiad.

Example 3 Given nine points in the space, among them there are no four points on a face. Join every pair of points and try to find the minimum n so that you can color everyone of any n line segments by red or blue arbitrarily. In the set of the n line segments, there must be a monochromatic triangle.

Solution By assumption there are no four points in a face, which assures that among nine points there are no three points on a line. So this problem is still a planar graph problem. The problem can be stated as follows: There are nine points on a plane and among them there are no three points on a line and there are 36 lines. How many lines should we take so that there must be a monochromatic triangle when we color the graph in two colors randomly?

We construct a graph G with 9 vertices, 32 edges and in two colors. We color the edges joining the vertex v_1 and v_2, v_3, v_8, v_9 by red (a solid line) while we color the edges joining the vertex v_1 and v_4, v_5, v_6, v_7 by blue (a dotted line). We divide the vertices other than v_1 into four groups: I : (v_2, v_3); II : (v_4, v_5); III : (v_6, v_7); IV : (v_8, v_9). We call I and II, II and III, III and IV adjacent groups. Except v_1, two vertices belonging to one group are not adjacent, two vertices belonging to two different adjacent groups are joined by a solid line (red) and two vertices belonging to two different groups which are not adjacent are joined by a dotted line (blue). Fig. 8.4 shows that there are $\binom{9}{2} - 4 = 32$ edges in the graph G which contains 16 red edges (solid lines) and 16 blue edges (dotted lines). It is not difficult to know that G contains no monochromatic triangle. So $n \geqslant 33$.

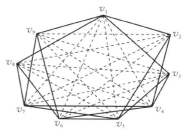

Fig. 8.4

Next let us prove $n \leqslant 33$. Suppose 33 edges connecting have been colored. While there are 3 edges which have not been colored. Without loss of generality, we denote the 3 edges by e_1, e_2, e_3. Choose one end v_1, v_2, v_3 from e_1, e_2, e_3, respectively. Then we delete the three vertices from K_9 and the remaining 6 vertices form a graph K_6. So if we color the graph by red and blue, the graph must contain a monochromatic triangle.

So $n = 33$.

To generalize Theorem 1, we first need to increase the number of colors.

We use k colors c_1, c_2, $\ldots$, c_k to color the complete graph K_n. We call the complete graph K_n k-color complete graph K_n if every edge is colored in one color. We can imagine if n is large enough, k-color complete graph K_n must contain monochromatic triangle. We denote the least n by r_k. In Theorem 1, $r_2 = 6$. It is clear that $r_1 = 3$.

The existence of r_k was firstly proved by British mathematician and logician Ramsey. We call r_k the *Ramsey number*. Concerning r_k we have the next conclusion.

Theorem 2 (1) For every positive integer k, the Ramsey number r_k exists. When $k \geqslant 2$,
$$r_k \leqslant k(r_{k-1} - 1) + 2.$$

(2) For any natural number k,
$$r_k \leqslant 1 + 1 + k + k(k-1) + \cdots + \frac{k!}{2!} + \frac{k!}{1!} + k!.$$

Proof (1) Apply induction on k. We know that r_1, r_2 exist and $r_1 = 3$, $r_2 = 6 \leqslant 2(r_1 - 1) + 2$.

Suppose that r_k exists and $r_k \leqslant k(r_{k-1} - 1) + 2$ holds. Take $n = (k+1)(r_k - 1) + 2$ and let K_n be a $(k+1)$-color complete graph whose vertices are A_1, A_2, ..., A_n. Take a vertex A_1 from K_n randomly, which is adjacent to $n - 1 = (k+1)(r_k - 1) + 1$ edges. There are $k+1$ colors in these edges. By the Pigeonhole Principle, in these edges there are at least r_k monochromatic edges. Suppose that these r_k edges are A_1A_2, A_1A_3, ..., $A_1A_{r_k+1}$, which are all colored in color c_1. Consider r_k-subset K_{r_k} consisting of the vertices A_2, A_3, ..., A_{r_k+1}. If K_{r_k} contains an edge with colored c_1 such as A_2A_3, then $\triangle A_1A_2A_3$ is a monochromatic triangle. If K_{r_k} contains no edges with color c_1, then there are k colors coloring the K_{r_k}, which means that K_{r_k} is k-color complete graph. By the induction hypothesis, K_{r_k} contains a monochromatic triangle.

In summary, K_n contains a monochromatic triangle. We know that $M = \{m \mid$ any $(k+1)$-color complete graph with m vertices that contains a monochromatic triangle$\}$ is a nonempty subset of the natural number set N, then there is a minimum r_{k+1} and
$$r_{k+1} \leqslant n = (k+1)(r_k - 1) + 2.$$

(2) Apply induction. When $k = 1$, $r_1 = 3 \leqslant 1 + 1 + 1$. Suppose that the property is true for k, then by (1) and the induction hypothesis, we get

$$r_{k+1} \leq (k+1)(r_k - 1) + 2$$
$$\leq (k+1)\left[1 + k + k(k-1) + \cdots + \frac{k!}{2!} + \frac{k!}{1!} + k!\right] + 2$$
$$= (k+1) + (k+1)k + (k+1)k(k-1) + \cdots$$
$$+ \frac{(k+1)!}{2!} + \frac{(k+1)!}{1!} + (k+1)! + 2$$
$$= 1 + 1 + (k+1) + (k+1)k + \cdots$$
$$+ \frac{(k+1)!}{2!} + \frac{(k+1)!}{1!} + (k+1)!.$$

So the property is also true for $k+1$.

If we make use of the expanding formula of base number of the natural logarithm in advanced mathematics

$$e = 1 + \frac{1}{1!} + \frac{1}{2!} + \cdots + \frac{1}{n!} + \cdots,$$

we can simplify (2) and get $r_k \leq [k!e] + 1$. Here we denote the largest integer no more than x by $[x]$.

Though Theorem 2 proves the existence of r_k and gives an upper bound of r_k, we only know three exact values of r_k. Other than $r_1 = 3$, $r_2 = 6$ mentioned above, we also know $r_3 = 17$. In face, according to (1) in Theorem 2, we know

$$r_3 \leq 3(r_2 - 1) + 2 = 3 \times 5 + 2 = 17.$$

We follow the conclusion of Theorem 2 and obtain a result which was set as a problem in the 6th International Mathematical Olympiad in 1964:

There are 17 scientists among which everyone communicates with any other person. They only discuss three problems when they are communicating and every two scientists can only discuss one problem. Prove that there are at least three scientists discussing one problem. On the other hand, we can color the complete graph K_{16} to using three colors so that the graph contains no monochromatic triangle. As Fig. 8.5

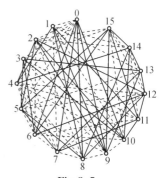

Fig. 8.5

shows us, a solid line represents a red edge, a dotted line represents a blue edge and no line represents a yellow edge. This means that $r_3 \geq 17$. So $r_3 = 17$.

We can have another generalization of Theorem 1.

Suppose every edge of the complete graph K_n is colored in red or blue, which means that K_n is a two-color (red, blue) complete graph K_n. For two constant natural numbers p, q, when n is large enough, the two-color (red, blue) complete graph K_n must contain a red K_p or a blue K_q. We denote the least n satisfying the above proposition by $r(p, q)$. We also call $r(p, q)$ the *Ramsey number*.

Using the concept of subgraph and complementary graph, we call $r(p, q)$ the minimum n so that any n-subgraph G of the complete graph K_n contains a complete subgraph K_p or its complementary graph $\overline{G}$ contains a complete subgraph K_q.

By definition and Theorem 1, we know that $r(3, 3) = r_2 = 6$. Furthermore, it is easy to find that $r(1, q) = r(p, 1) = 1$.

In order to understand the next general conclusion of $r(p, q)$, we prove an example: $r(3, 4) = 9$. First prove $r(3, 4) \leq 9$ and we can also give another example with the same form as Example 1.

Example 4 Prove that among any nine persons, you can find three persons knowing each other or four persons not knowing each other.

Proof We denote nine persons by nine vertices $A_1, A_2, \ldots, A_9$. We join every two vertices and if A_i knows A_j, then color $A_i A_j$ red. Otherwise, we color it blue. What we need prove is that in the two-color complete graph K_9 there must exist a red K_3 or a blue K_4.

If a vertex is adjacent to no less than four red edges, we denote them by $A_1 A_2$, $A_1 A_3$, $A_1 A_4$, $A_1 A_5$. If there is an edge joining A_2, A_3, A_4, A_5 by a red line, for example $A_2 A_3$. Then $\triangle A_1 A_2 A_3$ is a red triangle. If there is no edge join A_2, A_3, A_4, A_5 by a red line, then A_2, A_3, A_4, A_5 form a blue triangle. We complete the proof.

If the number of red edges adjacent to every vertices is less than 4, then the number of blue edges adjacent to every vertex is no less than 5. Consider the graph consists of A_1, A_2, ..., A_9 and all the blue edges. Since the number of odd edges is even, there must be an even vertex such as v_1. Since the number of blue edges adjacent to A_1 is even, the number of blue edges adjacent to A_1 is no less than 6. Let $A_1 A_2$, $A_1 A_3$, ..., $A_1 A_7$ be blue edges. Consider six vertices A_2, A_3, ..., A_7, every two of them are incident to be red or a blue edge. By Theorem 1, the graph contains a red triangle or a blue triangle. In the first case, the proposition is true, and we complete the proof. In the second case, let $\triangle A_2 A_3 A_4$ be a blue triangle. Then the complete graph K_4 with vertices A_1, A_2, A_3, A_4 is blue. The proposition is also true.

Considering a complete two-color complete graph K_8 as Fig. 8.6 shows us, we denote red edges by solid lines and blue edges by dotted lines. There exists a color of graph K_8 such that K_8 contains no red K_3 and no blue K_4. It means that $r(3, 4) > 8$.

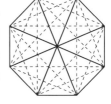

Fig. 8.6

In summary, $r(3, 4) = 9$.

Concerning $r(p, q)$, we have the next conclusion.

Theorem 3 (1) $r(2, q) = q$, $r(p, 2) = p$.

(2) $r(p, q) = r(q, p)$.

(3) When $p \geqslant 2$, $q \geqslant 2$,

$$r(p, q) \leqslant r(p-1, q) + r(p, q-1).$$

The inequality holds when $r(p-1, q)$ and $r(p, q-1)$ are all even.

We prove the theorem using subgraphs and their complementary graphs. The proof of (3) is very difficult. We may refer to the proof of Example 4.

Proof (1) Let G be a graph with q vertices. If two vertices in G are adjacent, G contains K_2. Otherwise $\overline{G}$ contains a K_q. So $r(2, q) \leqslant q$. Since $q-1$ vertices which are not adjacent to each other form a graph G. Obviously G contains no K_2 and its complementary graph

$\overline{G}$ contains no K_q. So $r(2, q) \geq q$.

In summary, $r(2, q) = q$. Similarly, we can prove $r(p, 2) = p$.

(2) If G contains $r(p, q)$ vertices, then $\overline{G}$ also contains $r(p, q)$ vertices. So $\overline{G}$ contains K_p or G contains K_q. In other words, G contains K_q or $\overline{G}$ contains K_p. So $r(p, q) \geq r(q, p)$. Similarly, $r(q, p) \geq r(p, q)$. So $r(p, q) = r(q, p)$.

(3) Suppose G contains $r(p, q-1) + r(p-1, q)$ vertices, and v_1 is a vertex of G.

If $d(v_1) \geq r(p-1, q)$, let $\delta = r(p-1, q)$ vertices be $v_2, v_3, \ldots, v_\delta, v_{\delta+1}$ which are adjacent to v_1. Then we remove other vertices and their edges from G to get G_1. According to the definition of $\delta = r(p-1, q)$, G_1 contains K_{p-1} or $\overline{G}_1$ contains K_q. If G_1 contains K_{p-1}, then in G, K_{p-1} and v_1 form a complete graph K_p. If $\overline{G}_1$ contains K_q, then $\overline{G}$ also contains this K_q.

If the number of the vertices adjacent to v_1 is less than $r(p-1, q)$, then v_1 is not adjacent to at least $r(p, q-1)$ vertices. We denote the vertices not adjacent to v_1 by $v_2, v_3, \ldots, v_s, v_{s+1}$, where $\varepsilon = r(p, q-1)$. We remove the vertices except $v_2, v_3, \ldots, v_{s+1}$ and their adjacent edges to get G_2. According to the definition of $\varepsilon = r(p, q-1)$, G_2 contains K_p or $\overline{G}_2$ contains K_{q-1}. If G_2 contains K_p, G also contains this K_p. If $\overline{G}_2$ contains K_{q-1}, then in $\overline{G}$, v_1 and K_{q-1} form a K_q.

In summary, we get

$$r(p, q) \leq r(p, q-1) + r(p-1, q).$$

If $r(p, q-1)$ and $r(p-1, q)$ are even, we choose a graph G with $r(p, q-1) + r(p-1, q) - 1$ vertices. Since the number of odd vertices is even and $r(p, q-1) + r(p-1, q) - 1$ is odd, G contains an even vertex v_1. For v_1, either $d(v_1) \geq r(p-1, q) - 1$, or v_1 is not adjacent to at least $r(p, q-1)$ vertices. Since the number of vertices adjacent to v_1 is even, in the first case, $d(v_1) \geq r(p-1, q)$. According to the same method we get

$$r(p, q) \leqslant r(p, q-1) + r(p-1, q) - 1$$
$$< r(p, q-1) + r(p-1, q).$$

We complete the proof of Theorem 3.

By Theorem 3, we get an upper bound of some Ramsey numbers $r(p, q)$. For example,

$$r(3, 3) \leqslant r(3, 2) + r(2, 3) = 3 + 3 = 6,$$
$$r(3, 4) \leqslant r(3, 3) + r(2, 4) - 1 \leqslant 6 + 4 - 1 = 9,$$
$$r(3, 5) \leqslant r(3, 4) + r(2, 5) \leqslant 9 + 5 = 14,$$
$$r(4, 4) \leqslant r(4, 3) + r(3, 4) = 9 + 9 = 18.$$

We have proved $r(3, 3) = 6$, $r(3, 4) = 9$, similarly, we can prove $r(3, 5) = 14$, $r(4, 4) = 18$. By a known inequality, of course we need only to prove $r(3, 5) > 13$, $r(4, 4) > 17$. Here, we only prove the first inequality and leave out the second proof.

Consider the graph which Fig. 8.7 shows us, it contains no K_3 and its complementary graph $\overline{G}$ contains no K_5. So $r(3, 5) > 13$. Applying Theorem 3, we can get a simple upper bound of $r(p, q)$, as shown in Theorem 4.

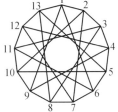

Fig. 8.7

Theorem 4 When $p \geqslant 2$, and $q \geqslant 2$,

$$r(p, q) \leqslant \binom{p+q-2}{p-1}.$$

Proof Note that $l = p + q$. We apply induction on l.

When $l = 4$, $p = q = 2$. The left side is $r(2, 2) = 2$ and the right side is $\binom{4-2}{1} = 2$.

Suppose the theorem is true when $l = k$ ($k \geqslant 4$). When $l = k+1$, consider the case when $p = k-1$, $q = 2$, or $p = 2$, $q = k-1$,

$$r(k-1, 2) = r(2, k-1) = k - 1 = \binom{k-1}{k-2} = \binom{k-1}{1}.$$

The theorem is true. When $p \geqslant 3$, and $q \geqslant 3$, we have $p + q = k + 1$, apply (3) in Theorem 3 and by induction, we have

$$r(p,q) \leq r(p-1,q) + r(p,q-1)$$
$$\leq \binom{p+q-3}{p-2} + \binom{p+q-3}{p-1}$$
$$= \binom{p+q-2}{p-1}.$$

The theorem is also true. By induction, Theorem 4 is proved.

In spite of the result, it is difficult to find the exact value of $r(p, q)$. Except for $r(1,q) = r(p,1) = 1$, $r(p,2) = p$, $r(2,q) = q$, we know only a few values of $r(p,q)$. They are listed in the next table in which the numerator represents the low bound and the denominator represents the upper bound.

p \ q	3	4	5	6	7	8
3	6	9	14	18	23	$\frac{28}{29}$
4	9	18	$\frac{25}{28}$	$\frac{34}{36}$		
5	14	$\frac{25}{28}$	$\frac{42}{55}$	$\frac{51}{94}$		
6	18	$\frac{34}{36}$	$\frac{38}{94}$	$\frac{102}{178}$		

If we combine the above two generalizations, we obtain the next generalization.

Color the complete graph K_n in l colors $c_1, c_2, \ldots, c_l$. Every edge can be colored in only one color and we get an l-color complete graph. When n is large enough, l-color complete graph K_n contains a c_1-color complete subgraph K_{p_1} or a c_2-color complete subgraph $K_{p_2}, \ldots$ or a c_l-color complete subgraph K_{p_l}. We denote the minimum n satisfying the above propositions by $r(p_1, p_2, \ldots, p_l)$. We also call it a Ramsey number.

If we introduce the definition of hypergraph, we can make

further generalization. We do not elaborate here.

Example 5 We divide 1, 2, 3, 4, 5 into two groups A, B randomly. Prove that it is possible to find two numbers in a group and the difference of the two numbers is the same as one number of this group.

Proof We divide 1, 2, 3, 4, 5 into two groups A, B randomly. Choose six vertices and label them as 1, 2, 3, 4, 5, 6. For any two vertices $i > j$, it is always true that $1 \leqslant i - j \leqslant 5$. For the two vertices $i > j$, if $i - j$ is in group A, we color the edge ij red; if $i - j$ is in group B, we color the edge ij blue. So we get a 2-color complete graph K_6. By Example 1, this K_6 contains a monochromatic triangle which is $\triangle ijk$ ($i > j > k$). This means that $a = i - k$, $b = i - j$, $c = j - k$. The three numbers are in one group, and

$$a - b = (i - k) - (i - j) = j - k = c.$$

We have completed the proof.

Remark In this example, it is possible that $b = c$, then $a = 2b$. The problem can be rewritten as follows. We divide 1, 2, 3, 4, 5 into two groups A, B randomly. Prove that it is possible to find a number in a group so that it is twice one number in the group or the sum of two numbers in the same group.

Question 8 in Exercise 8 of this chapter is an IMO problem in 1978 which is an extension or generalization of this example. Generalize Question 8 further, we get the famous Schur Theorem (Question 7).

A variate of monochromatic triangle is heterochromous triangle whose three edges are colored in three distinct colors. The following is a question from Hungarian Mathematical Olympiad.

Example 6 There are $3n + 1$ persons in a club. Any two persons can play one of the three games: Chinese chess, the game of go, Chinese checkers. It is known that everyone must play Chinese chess with n persons, the game of go with n persons and Chinese checkers with n persons. Prove that among the $3n + 1$ persons, there must be three persons so that there are Chinese chess player, go player, and

Chinese checkers player among the three.

Proof We denote $3n+1$ persons by $3n+1$ vertices. If two persons play Chinese chess, the corresponding edge of them is colored red. If two persons play the game of go, the corresponding edge of them is colored blue. If two persons play Chinese checkers, the corresponding edge of them is colored black. Then we get a 3-color complete graph K_{3n+1}. What we must prove is that in this 3-color complete graph K_{3n+1}, there must be a heterochromous triangle.

If two edges adjacent to one vertex are not monochromatic, we call the angle of the two edges heterochromous angle. A triangle is heterochromous if and only if its three angles are heterochromous angles. Every vertex is adjacent to $3n$ edges during which there are n red edges, n blue edges, n black edges, respectively. Therefore, the number of the heterochromous angles induced by one vertex is $\binom{3}{2}n^2 = 3n^2$. The 3-color complete graph K_{3n+1} contains $3n^2(3n+1)$ heterochromous angles at all. On the other hand, complete graph K_{3n+1} contain $\binom{3n+1}{3} = \frac{1}{2}n(3n+1)(3n-1)$ triangles. We can regard these triangles as holes and heterochromous angles as pigeons. Since $3n^2(3n+1) > n(3n+1)(3n-1)$, the number of heterochromous angles in the 3-color complete graph K_{3n+1} is twice more than the number of triangles. By the Pigeonhole Principle, there must be a triangle with three heterochromous angles. We call the triangle a heterochromous triangle.

We often find the similar problems about the Ramsey problem in mathematics contest. We give some more examples to finish this chapter.

Example 7 There are 100 guests in a hall. Everyone of them knows at least 67 persons. Prove that among these guests you can find 4 persons any two of them know each other. (Polish Mathematical Competition in 1966)

Proof We denote the guests by 100 vertices $A_1, A_2, \ldots, A_{100}$. Join every two vertices and color it in red or blue. The edge joining A_i and A_j is colored in red if and only if A_i and A_j know each other. We use the language of graph theory to re-phrase this problem: In a red-blue two color complete graph K_{100}, if the number of red edges going out of every vertex is at least 67, then K_{100} contains a red complete subgraph K_4.

Take one vertex A_1. The number of red edges induced by it is no less than 67, so there must exist a red edge A_1A_2. Since the number of red edges induced by A_2 is no less than 67, the number of blue edges induced by A_1 and A_2 is at most $32 \times 2 = 64$. They involve 66 vertices and there must exist one vertex, for example A_3 so that A_1A_3 and A_2A_3 are red edges. The number of blue edges induced by A_1, A_2, A_3 is at most $32 \times 3 = 96$ and these blue edges involve 99 vertices. There must exist one vertex A_4 so that A_1A_4, A_2A_4, A_3A_4 are red edges. So the complete subgraph K_4 with vertices A_1, A_2, A_3, A_4 is red.

Example 8 We use pentagons $A_1A_2A_3A_4A_5$ and $B_1B_2B_3B_4B_5$ as the top and bottom faces of a prism. Every edge and every line segment A_iB_j, where i, $j = 1, 2, \ldots, 5$, are colored in red or blue. Every triangle which uses a vertex of the prism as its vertex and a line segment which has been colored as its edge is not a monochromatic triangle. Prove that the ten edges in the top and bottom faces are colored the same color. (The 21th IMO)

Proof First we prove that the five edges on the top face are colored the same color. Otherwise, there are at least two edges in the pentagon which are not monochromatic. So there are two adjacent edges such as A_1A_2, A_1A_5 which are not monochromatic. Without loss of generality, we suppose that A_1A_2 is red and A_1A_5 is blue. Among the edges joining A_1 and B_1, B_2, B_3, B_4, B_5 there are at least 3 monochromatic edges. Let A_1B_i, A_1B_j, A_1B_k be red edges (i, j, k are distinct). Since $\triangle A_1B_iB_j$ is not monochromatic, B_iB_j is a blue edge. Similarly, A_2B_i is also a blue edge. We can also know that A_2B_j is a red edge. Then $\triangle A_1A_2B_j$ is a red triangle. It is a contradiction.

Similarly, we can prove that the five edges on the bottom face are also monochromatic.

If the edges in the top and bottom face are not monochromatic.

We suppose that $A_1A_2A_3A_4A_5$ is red and $B_1B_2B_3B_4B_5$ is blue. Without loss of generality, let A_1B_1 be a blue edge. By the assumption that every triangle is not monochromatic, we can know that A_1B_5 and A_1B_2 are all red edges. So A_2B_2 and A_5B_5 are blue edges. Similarly, A_5B_1, A_5B_4, A_2B_1, A_2B_3, ... are all red edges and A_3B_3, A_4B_4 are blue edges. So A_4B_1 and A_4B_2 are blue edges and we can get a blue triangle $\triangle A_4B_1B_2$. It is a contradiction.

So the ten edges in the top and bottom faces are monochromatic.

Example 9 There are two international airlines X and Y serving 10 districts. For any two districts, there is only one company providing a direct flight (to and fro). Prove that there must be a company which can provide two tour routes so that the two routes do not pass through the same districts and each route passes through an odd number of districts.

Proof We denote the 10 districts by 10 vertices u_1, u_2, ..., u_{10}. If the flight between u_i and u_j is provided by X, then we join u_i and u_j by a red edge (a solid line); If the flight between u_i and u_j is provided by Y, then we join u_i and u_j by a blue edge (a dotted line). Then we can get a 2-color complete graph K_{10}. In order to prove the conclusion, it suffices to prove that there must be two monochromatic triangles or polygons having no common edge and an odd number of edges in K_{10}.

The 2-color complete graph K_{10} contains a monochromatic triangle. Let $\triangle u_8u_9u_{10}$ be a monochromatic triangle. By Example 1, we can know that the triangles constructed by the vertices u_1, u_2, ..., u_7 must contain a monochromatic triangle. Let $\triangle u_5u_6u_7$ be a monochromatic triangle. If the color of $\triangle u_5u_6u_7$ is the same as that of $\triangle u_8u_9u_{10}$, the conclusion holds. Then let $\triangle u_5u_6u_7$ be red and $\triangle u_8u_9u_{10}$ be blue.

The number of edges joining the vertex sets $\{u_5, u_6, u_7\}$ and $\{u_8, u_9, u_{10}\}$ is $3 \times 3 = 9$. By the Pigeonhole Principle, there must be five monochromatic edges. Let them be red edges. The five edges are induced by $\{u_8, u_9, u_{10}\}$, so there must exist a vertex which is adjacent to two red edges which are u_8u_6, u_8u_7. As Fig. 8.8 shows, there must also be another red triangle $\triangle u_6u_7u_8$.

Consider the 2-color complete graph K_5 whose vertices are u_1, u_2, u_3, u_4, u_5. If the K_5 contains a monochromatic triangle, whatever color the triangle is, together with the red triangle $\triangle u_6u_7u_8$ or the blue triangle $\triangle u_8u_9u_{10}$. K_{10} contains two monochromatic triangles with common edge and the same color. Otherwise, the 2-color complete graph K_5 contains no monochromatic triangle. It is easy to know that K_5 contains two monochromatic pentagons which are one red and one blue. We complete the proof.

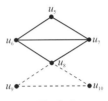

Fig. 8.8

Remark If we replace 10 districts by 9 districts, the conclusion is false. An example is given as follows. We divide 9 districts into 3 groups, i.e. $\{u_1, u_2, u_3, u_4, u_5\} = A$, $\{u_6, u_7, u_8\} = B$, $\{u_9\} = C$. The flights among the five districts in A are provided by X. The flights among the three districts in B are provided by Y. The flights between A and B, u_9 and A are provided by Y. The flights between u_9 and A are provided by X.

Exercise 8

1 In the space, there are six points. Join every two of them and color the lines in red or blue. Prove that there must be two monochromatic triangles.

2 In the space, there are eight points. Join every two of them and color the lines in two colors. Prove that there must exist three monochromatic lines which contain no common point.

3 In the space, there are six points. Any three points are not the vertices of an equilateral triangle. Prove that among these triangles, there is one triangle whose shortest side is also the longest line of another triangle.

4 Join nine distinct points on a circle to get 36 lines and color them in red or blue. Suppose any triangle with three vertices coming from the nine points contains a red line. Prove that there are four points and any edge joining two of them is red.

5 Prove that among any 19 persons, there must be 3 persons who know each other or 6 persons who do not know each other.

6 Prove that among any 18 persons, there must be 4 persons who know each other or do not know each other.

7 We divide the natural numbers $1, 2, \ldots, N$ into n groups. When N is large enough, there must be a group which contains x, y and their difference $|x - y|$. (Schur Theorem)

8 There are 1978 members in an international corporation. They come from 6 countries. We label them as $1, 2, \ldots, 1978$. Prove that there must be at least one member whose number is twice the number of his one fellow-country or the sum of two fellow-countrymen.

9 Prove that in a 2-color complete graph K_7, there must be two monochromatic triangles with no common edges.

10 In the space, there are six lines. Among them every three lines do not lie on a plane. Prove that there must be three lines satisfying one of the following three conditions: (i) Any two of them do not lie on a plane. (ii) Any two of them is parallel to each other. (iii) They meet at one point.

11 Find the minimum positive integer n so that any given n irrational numbers always contain three irrational numbers among which the sum of any two is also an irrational number.

12 Find the minimum positive integer n so that when the K_n is colored by two colors arbitrarily, there must be two monochromatic triangles which are colored by one color but contain no common edges.

13 In a football league, there are 20 football teams. In the first round, they are divided into 10 matches. In the second round, they are also divided into 10 matches. (Notice that the opponent of every team in different rounds can be the same.) Prove that before the third round, you can find 10 teams which have not played with each other.

Chapter 9 Tournament

In Chapter 1, we have said that the graph is a tool to describe the special relationship of some objects. The graph in the above chapter is an undirected graph. The relationship they describe is a symmetrical relationship. In daily life, many relationships are not symmetrical such as the relationship of knowing each other. When X knows Y, it does not mean that Y knows X. So is the relationship of winning or losing in a match. So we can have a new definition of directed graph.

We call a graph *directed* graph if we assign to every edge of the graph a direction. We call the edge of a directed graph an *arc*. If there is an arc joining the vertices v_i and v_j and the arrow of the arc points from v_i to v_j, we denote it by (v_i, v_j) and call v_i the starting point and call v_j the end point. Generally, we denote the directed graph by $D = (V, U)$. Here we denote the vertex set of D by V and the arc set of D by U. Fig. 9.1 shows us a directed graph. The vertex set is

$$V = \{v_1, v_2, v_3, v_4, v_5, v_6\},$$

and the arc set is

$$U = \{(v_1, v_2), (v_2, v_3), (v_5, v_2), (v_4, v_2), (v_4, v_6),\\ (v_5, v_6), (v_5, v_4), (v_3, v_5), (v_4, v_5)\}.$$

The directed graph in this chapter is also a simple directed graph, which is a graph without loops (an arc which starts and ends at the same point) and without multi-arcs (there are more

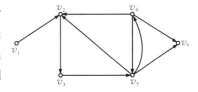

Fig. 9.1

than one arc joining v_i and v_j).

We say that v_i and v_j are adjacent if there is an arc (v_i, v_j) or (v_j, v_i) in the arc set of the directed graph G. Otherwise, we say that v_i and v_j are not adjacent. We call the number of the arcs whose starting points are v_i an *outdegree* of v_i, which is denoted by $d^+(v_i)$. We call the number of the arcs whose end points are v_i an *indegree* of v_i, which is denoted by $d^-(v_i)$.

We call a directed graph *tournament* graph if the graph contains n vertices and there is only an arc joining every two vertices. We denote the directed graph by $\overline{K}_n$.

Theorem 1 Let $v_1, v_2, \ldots, v_n$ be the vertices of tournament $\overline{K}_n$. Then

$$\begin{aligned} & d^+(v_1) + d^+(v_2) + \cdots + d^+(v_n) \\ &= d^-(v_1) + d^-(v_2) + \cdots + d^-(v_n) \\ &= \frac{1}{2}n(n-1). \end{aligned}$$

Proof Since every arc of $\overline{K}_n$ induces one indegree and one outdegree and there is only one arc joining every two vertices, the sum of indegree of every vertices in $\overline{K}_n$ is the same as the sum of outdegree of every vertices.

$$\begin{aligned} & d^+(v_1) + d^+(v_2) + \cdots + d^+(v_n) \\ &= d^-(v_1) + d^-(v_2) + \cdots + d^-(v_n) \\ &= \frac{1}{2}n(n-1). \end{aligned}$$

Example 1 n players $P_1, P_2, \ldots, P_n$ ($n > 1$) take part in a round robin. Every player plays only one game with any of other $n-1$ players. Suppose that there is no tie in the result and we denote the number of win and lose of P_r by w_r and l_r, respectively. Prove that

$$w_1^2 + w_2^2 + \cdots + w_n^2 = l_1^2 + l_2^2 + \cdots + l_n^2.$$

(The 26th American Putnam Mathematical Competition)

Solution Draw a tournament $\overline{K}_n$. We denote the person P_r by the vertex v_r. If P_i defeats P_j, we join v_i and v_j to get an arc (v_i, v_j). So w_r and l_r are the indegree and outdegree of v_r respectively. By Theorem 1,

$$w_1 + w_2 + \cdots + w_n = l_1 + l_2 + \cdots + l_n.$$

Note that $w_i + l_i = n - 1$ $(1 \leqslant i \leqslant n)$,

$$\begin{aligned}
&w_1^2 + w_2^2 + \cdots + w_n^2 - (l_1^2 + l_2^2 + \cdots + l_n^2) \\
&= (w_1^2 - l_1^2) + (w_2^2 - l_2^2) + \cdots + (w_n^2 - l_n^2) \\
&= (w_1 + l_1)(w_1 - l_1) + (w_2 + l_2)(w_2 - l_2) + \cdots + (w_n + l_n)(w_n - l_n) \\
&= (n - 1)[(w_1 + w_2 + \cdots + w_n) - (l_1 + l_2 + \cdots + l_n)] = 0.
\end{aligned}$$

So

$$w_1^2 + w_2^2 + \cdots + w_n^2 = l_1^2 + l_2^2 + \cdots + l_n^2.$$

In a directed graph $D = (v, u)$, there exists a sequence of distinct arcs $u_1, u_2, \ldots, u_n$. If the starting point of u_i is v_i and the end point of u_i is v_{i+1} $(i = 1, 2, \ldots, n)$. We call n the length of the directed path. v_1 is the starting point of the path and v_{n+1} is the end point. If $v_1 = v_{n+1}$, we call the path a *circuit*.

Example 2 The MO space city consists of 99 space stations. Any two stations are connected by a channel. Among these channels there are 99 two-way channels and others are one-way channels. If four space stations can be arrived at from one to another, we call the set of four space stations a connected four-station group.

Design a scheme for the space city so that we get the maximum number of connected four-station groups. (Find the exact number and prove your conclusion.) (The 14th China Mathematical Olympiad)

Solution We call an unconnected four-station group a bad four-station group. A bad four-station group has three possible situations:

(1) Station A has three channels AB, AC, AD which all leave A.

(2) Station A has three channels which all arrive at A.

(3) Stations A and B, stations C and D have two-way channels but the channels AC, AD all leave A, and BC, BD all leave B.

We denote all the bad four-station groups in (1) by S and others by T. Let us calculate $|S|$.

Since the space city contains $\binom{99}{2} - 99 = 99 \times 48$ one-way channels. We denote the number of channels leaving the i-th station by S_i. So

$$\sum_{i=1}^{99} S_i = 99 \times 48.$$

Now the number of bad four-station group in (1) which contains three channels induced by A is $\binom{S_i}{3}$. So

$$|S| = \sum_{i=1}^{99} \binom{S_i}{3} \geqslant 99 \times \binom{48}{3}.$$

The above inequality holds because $\binom{x}{3} = \frac{1}{6} x(x-1)(x-2)$ for $x \geqslant 3$ is a convex function. Since the number of all four-station groups is $\binom{99}{4}$, so the number of connected four-station groups is no more than

$$\binom{99}{4} - |S| \leqslant \binom{99}{4} - 99\binom{48}{3}.$$

Then we give an example so that the number of the connected four-station groups is $\binom{99}{4} - 99\binom{48}{3}$. Let the number of the channels from and to every station A_i be both 48. Every station has two two-way channels and there are only type S groups of bad four-stations and no type T groups of bad four-stations.

We put 99 stations on the vertices of an inscribed polygon with 99 sides and assume the longest diagonals of the regular polygon with 99 sides all two-way channels. So every vertex is adjacent to two two-way channels. For station A_i, there are one-way channels leaving A_i and joining the next 48 stations in the clockwise direction, and one-way channels arriving at A_i and joining the next 48 stations in the counterclockwise stations. Then we will prove there are only type S groups of bad four-station groups and no T groups of bad four-station groups.

Suppose $\{A, B, C, D\}$ is a four-station group.

(i) If there are two two-way channels in the four-station group, clearly they are connected.

(ii) If there are only one two-way channel AC in the four-station group, each of B and D forms a cycle with A, C. Of course, they are connected.

(iii) So if the bad four-station group contains no two-way channel, it can only be one of (1) and (2). If it is (2), without loss of generality, we suppose that the 3 channels of A all arrive at A and B, C, D all come from the next 48 stations of A in the clockwise direction. Let D be the farthest station from A. So AD, BD, CD all leave from D, which means that all the bad four-station groups are of the type S group.

In summary, there are at most $\binom{99}{4} - 99\binom{48}{3}$ connected four-station groups.

Theorem 2 There exists a vertex in a tournament so that there is a path from it to any other vertices. The maximum length of the paths is 2.

Proof Suppose that the vertex with the maximum outdegree of a tournament $\overline{K}_n$ is v_1. We denote the end point set of the arcs whose starting point is v_1 by $N^+(v_1)$. If the conclusion is false, there must be a vertex v_2 ($v_2 \neq v_1$), where $v_2 \notin N^+(v_1)$. For every vertex $u \in N^+(v_1)$, there is one arc (v_2, u) from v_2 to u together with the arc (v_2, v_1). So $d^+(v_2) \geq d^+(v_1) + 1$, which contradicts the fact that degree of v_1 is maximum. The proof is complete.

Example 3 Every athlete who takes part in the single round robin must play one game with any other athlete and there is no tie. Prove that among these players, you can find such athletes that the persons who were defeated by him and the persons who were defeated by the person he defeats can contain all other athletes. (Hungarian mathematic contest)

Use tournament on the round robin, which is Theorem 2. We omit the proof.

Example 4 n ($n \geqslant 3$) athletes take part in a single round robin and use the result to find good athletes. The requirement that A is selected to be a good athlete is that for any other athlete B, either A defeats B or there exists C so that C defeats B and A defeats B. If only one athlete meets the above requirement, show that he defeats any other athletes.

Solution We denote n athletes by n vertices. If v_i defeats v_j, we draw an arc from v_i to v_j and get a tournament $\overline{K}_n$. Without loss of generality, we suppose that the outdegree of v_1 is maximum in the $\overline{K}_n$, according to Theorem 2, v_1 is a good athlete. What we will prove is that v_1 can arrive at any other vertex by a path whose length is 1, which means that the indegree of v_1 is $d^-(v_1) = 0$.

Suppose that the proposition is false. We denote the set of arcs with starting point v_1 by $N^-(v_1) = \{v_{i_1}, v_{i_2}, \ldots, v_{i_r}\}$, $r \geqslant 1$. Consider the $\overline{K}_r$ consisting of $v_{i_1}, v_{i_2}, \ldots, v_{i_r}$. We suppose that the outdegree of v_{i_1} is maximum in $\overline{K}_r$. By Theorem 2, the length of path from v_{i_1} to each of $v_{i_2}, \ldots, v_{i_r}$ is no more than 2. Since v_1 can arrive at other vertices except $\{v_{i_1}, v_{i_2}, \ldots, v_{i_r}, v_{i_1}\}$, then v_{i_1} can arrive at vertices except $v_{i_1}, v_{i_2}, \ldots, v_{i_r}$ through the paths whose length is no more than 2. Therefore, in the tournament $\overline{K}_r$, v_{i_1} can arrive at any other vertex through a path whose length is no more than 2. Hence v_{i_1} is also a good athlete, which contradicts the fact that v_1 is the only good athlete. So $N^-(v_1) = \emptyset$ or $d^-(v_1) = 0$. We have completed the proof.

Remark This problem gives a property of the tournament $\overline{K}_n$: If the vertex of $\overline{K}_n$ with the maximum outdegree is unique, then the outdegree of this vertex is $n - 1$.

Theorem 3 Tournament $\overline{K}_n$ contains a Hamiltonian path whose length is $n-1$.

Proof Apply induction on the number of vertices n. When $n = 2$, clearly the proposition is true.

Suppose the proposition is true for $n \leqslant k$. When $n = k+1$, from $k+1$ vertices we take a vertex v. Remove v and the arcs adjacent to v from $\overline{K}_{k+1}$. By induction, $\overline{K}_{k+1} - v$ contains a Hamiltonian path v_1, $v_2, \ldots, v_k$.

If there is an arc (v_k, v), then $v_1, v_2, \ldots, v_k, v$ is a Hamiltonian path. If there is an arc (v, v_1), then $v, v_1, v_2, \ldots, v_k$ is also a Hamiltonian path.

Otherwise there exist arc (v, v_k) and (v_1, v). Then there must be an i $(1 \leqslant i \leqslant k-1)$ so that the arcs (v_i, v), (v, v_{i+1}) both exist. Now $v_1, \ldots, v_i, v, v_{i+1}, \ldots, v_k$ is a Hamiltonian path as Fig. 9.2 shows us.

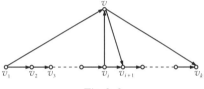

Fig. 9.2

Example 5 In a match of Chinese chess, every two players should play a game. Prove that we can label the players so that every player is not defeated by the player whose number follows immediately after his.

Solution Suppose there are n players. We denote n players by n vertices $v_1, v_2, \ldots, v_n$. When v_i is not defeated by v_j, we draw an arc from v_i to v_j to get (v_i, v_j). Then we get a tournament $\overline{K}_n$. By Theorem 3, $\overline{K}_n$ contains a Hamiltonian path, so we can label them according to the order of the path.

Theorem 4 The tournament $\overline{K}_n (n \geqslant 3)$ contains a circuit which is a triangle, if and only if there are two vertices v and v' satisfying

$$d^+(v) = d^+(v').$$

Proof Let vertices v and v' satisfy $d^+(v) = d^+(v')$. We will prove that $\overline{K}_n$ contains a circuit which is a triangle.

Without loss of generality, we assume that there is an arc (v, v')

and draw arcs from v' to everyone of v_1, v_2, ..., v_k where ($k = d^+(v)$). Then there must be v_j ($1 \leq j \leq k$) so that there is an arc from v_j to v. Otherwise, $d^+(v) \geq k+1 > d^+(v')$ and v, v', v_j form a triangle. We have proved the sufficient condition.

If the outdegree of every vertex of $\overline{K}_n$ is different, we can prove by induction that $\overline{K}_n$ contains no triangular circuit. When $n = 3$, it is easy to see that the outdegree of a triangle is 0, 1, 2 and the triangle cannot form a circuit.

Suppose the proposition is true for $n = k$. Consider the tournament $\overline{K}_{n+1}$. If the outdegree of every vertex is different, they are 0, 1, 2, $\cdots$, k in order. Suppose $d^+(v') = k$. Remove the vertex v' and its adjacent arcs. By induction hypothesis, $\overline{K}_k - v'$ contains no triangular circuit. Clearly $\overline{K}_{k+1}$ contains no triangular circuit. We have proved the necessary condition.

Exercise 9

1 Among n ($n > 4$) cities, every two cities have a path to join them. Prove that we can change these paths to one-way path so that it is possible to go from one city to another city through at most one other city.

2 If a tournament $\overline{K}_n$ contains a circuit, then $\overline{K}_n$ contains a trianglular circuit.

3 In a country, N cities are connected by air routes. For any route, the airplanes can fly only along one direction. An air route satisfies the condition f: Any plane which starts from one city cannot return to the same city. Prove that it is possible to design an airline system so that every two cities are connected by an air route and the system also satisfies condition f.

4 In a volleyball round robin, if team A defeats team B or team A defeats team C and team C defeats team B, we say that A is

superior to B and we also call the team superior to any other team the champion. According to this regulation, can two teams both win the championship?

5 n players take part in a match in which everyone plays with several other players. Suppose that there is no tie in a game. If the result that v_1 defeats v_2, v_2 defeats v_3, ..., v_k defeats v_1 does not appear. Prove that there must be a player who wins all games and another player who loses all games.

6 If among n persons v_1, v_2, ..., v_n every two persons v_i and v_j have one ancestry v_k. Everyone can be an ancestry of oneself. Prove that the n persons have a common ancestry.

7 A, B, C, D play table tennis and every two persons should play against each other. At last, A defeats D and A, B, C win the same number of games. How many games does C win?

8 n $(n \geqslant 3)$ players take part in a round robin. Every pair should play a game and there is no tie. There is no player who defeats all other players. Prove that among them there must be three persons A, B, C so that A defeats B, B defeats C and C defeats A.

9 There are 100 species of insects. Among every two of them there is one species who can eliminate another species. (But A eliminates B, B eliminates C, which does not mean that A eliminates C.) Prove that the 100 species of insects can be arranged in an order so that any species can perish another species next to it.

Solutions

Exercise 1

1 The graph G is shown in Fig. 1.

2 In a simple graph, every edge is adjacent to two distinct vertices and among every two vertices there is at most one edge joining them. Now there are n vertices, so the number of edges is at most $\binom{n}{2} = \dfrac{n(n-1)}{2}$.

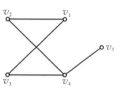

Fig. 1

Since not all simple graphs are complete graphs, then $e \leqslant \dfrac{n(n-1)}{2}$.

3 As shown in Fig. 2, we establish the relationship of the vertices: $v_1 \leftrightarrow u_1$, $v_2 \leftrightarrow u_2$, $v_3 \leftrightarrow u_3$, $v_4 \leftrightarrow u_4$, $v_5 \leftrightarrow u_5$ and the relationship of the edges: $e_1 \leftrightarrow e'_1$, $e_2 \leftrightarrow e'_2$, $e_3 \leftrightarrow e'_3$, $\ldots$, $e_8 \leftrightarrow e'_8$. The number of the vertices and that of the edges of the two graphs are equal. So we can establish a corresponding relationship between the vertices and edges of the two graphs. So the two graphs are isomorphic.

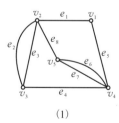 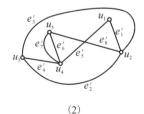

(1)   (2)

Fig. 2

4 Construct a graph as follows. We denote a medicine-chest by a vertex. We denote every two medicine-chests v_i and v_j which contain one common medicine by the edges (v_i, v_j). By the hypothesis, the graph is a complete graph K_n and the number of the kinds of the medicine is equal to the number of the edges $\frac{1}{2}n(n-1)$.

5 We denote n professors by n vertices $v_1, v_2, \ldots, v_n$ and join by an edge if the two corresponding persons know each other. We divide the n persons into two groups randomly. There are finite ways to divide them. Conside the number of edges joining the vertices of the two groups. There must be a division so that S is the largest. Now $d_i \geq d'_i$ ($i = 1, 2, \ldots, n$). Otherwise, if $d_1 < d'_1$, we transfer v_1 from a group to the other group. The number of S increases by $d'_1 - d_1 > 0$, which contradicts the fact that S is the largest.

6 Team A had played 8 matches with 8 teams and did not play with the other 9 teams. Suppose the 9 teams had played with each other in 8 rounds. Since every team had played 8 games, the 9 teams had not played with other teams. But the 9 teams could only play 4 games, so there must be one team which had played with other teams. A contradiction. So among the 9 teams there must be two teams B and C which had not played with each other. Then A, B, C had not played with each other.

7 We denote n delegates by n vertices. If two delegates have shaken their hands, we join the corresponding vertices and get the graph G. If among any four vertices v_1, v_2, v_3, v_4 in G, every vertex has its adjacent vertex, we denote them by v'_1, v'_2, v'_3, v'_4. By the known condition, among v_1, v_2, v_3, v_4 there is a vertex v_1 which is not adjacent to the other three vertices v_2, v_3, v_4. So $v'_1 \neq v_2, v_3, v_4$. If $v'_2 \neq v'_1$, among four vertices v_2, v_3, v'_1, v'_2, there is no vertex which is adjacent to the other three vertices. So $v'_2 = v'_1$. Similarly, $v'_3 = v'_1$. Among four vertices v_1, v_2, v_3, v'_1, there is no vertex which is adjacent to any other vertex. So among any 4 vertices there must be one vertex which is not adjacent to the other $n-1$ vertices.

8 We denote these students by $3n$ vertices. We divide the students from three schools into three vertex set which is denoted by X, Y and Z. If u and v are from different schools and they know each other, then join the vertices representing them to get a graph G. Suppose $x \in X$, we denote the number of vertices which are adjacent to x and lie in Y and Z by k and l respectively. Then $k + l = n + 1$. We denote the maximum number between k and l by $m(x)$. Let x go through every in X and the maximum value of $m(x)$ is denoted by m_X. Similarly, we definite m_Y and m_Z. We denote the maximum of m_X, m_Y and m_Z by m. Without loss of generality, suppose that $m = m_X$ and $x_0 \in X$ so that $|Y_1| = m$. ($|Y_1|$ is the number of the vertices in Y adjacent to x_0.) The number of vertices in Z adjacent to x_0 is $n + 1 - m \geq 1$. Suppose $z_0 \in Z$ is adjacent to x_0. If there is $y_0 \in Y_1$ adjacent to z_0, then $\triangle x_0 y_0 z_0$ is a triangle in G. If every vertex y in Y_1 is not adjacent to z_0, the number of vertices in Y adjacent to z_0 is no more than $n - m$. So the number of vertices in X adjacent to z_0 is no less than $n + 1 - (n - m) = m + 1$, which contradicts the fact that m is maximum. Then we prove G must contain $\triangle x_0 y_0 z_0$. For a diagram, see Fig. 3.

9 When $n = 1$, there are two red squares which are adjacent. Clearly, it is a rectangle. Suppose it is true when $n \leq k$, which means we can divide $2k$ connected squares into k rectangles. When $n = k + 1$,

(i) For $2k + 2$ squares, if we remove a pair of adjacent red squares from the graph, the graph is also connected. By induction, the conclusion is true.

(ii) If we remove a pair of adjacent red squares from the graph, the graph becomes several connected subgraphs. The number of red squares in every subgraph is even. By induction, the conclusion is true.

(iii) If we remove a pair of adjacent red squares from the graph, there is a connected subgraph in which there are odd red

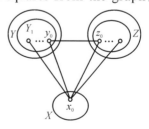

Fig. 3

squares:

When $n = 2$, there is a graph in the shape of the letter T with two rectangles 1×3 and 1×1 satisfying the requirement.

When $n \neq 2$, consider the T-shaped graph at the top-left corner of all the squares. After removing these two squares, we get at most two connected graphs.

If the number of the red squares of the two connected subgraphs is odd after removing the two squares at the top-left corner, then remove the two rectangles 1×3 and 1×1. It is easy to know they are also connected and the number of red squares is even. In summary, when $n = k + 1$, the conclusion is true. So the conclusion is true.

10 If 2000 members know each other, then the number of all the members in the delegation is 2000. Therefore, we suppose that two members u and v do not know each other. We prove in three steps.

(i) Any two members other than u and v know each other. Suppose that a and b are another the two members. By the hypothesis, among the four members a, b, u, v, there must be a person who knows another three persons. The person must be a or b, which means a and b know each other.

(ii) If u and v know every one of the remaining 1998 persons, then in the delegation there are 1998 persons who know all the other persons. Suppose that a is any member other than u, v. By the hypothesis, a knows u, v. Suppose that b is another member. By the above proof, a and b know each other. By the fact that b is arbitrary, a knows every other member. Also by the fact that a is arbitrary, the 1998 members in delegation other than u and v know all other members.

(iii) If one of u, v does not know all other 1998 members, then there are 1997 members knowing all the members. Suppose, other than v, u does not know another member w. Let a be any one of 1997 member other than u, v, w. By the hypothesis, among a, u, v, w there is only one person a who knows the other three. It means that each of u, v, w knows the other 1997 persons of the delegation.

114   Graph Theory

In summary, the number of persons who know all members of the delegation is at least 1997.

11 By assumption, everyone has friends. Suppose that k ($k \leq m$) persons are friends. Since they have a common friend, $k+1$ persons are friends. Similarly, we can deduce that $m+1$ persons $A_1, A_2, \ldots, A_{m+1}$ are friends. Next we will prove that there are no other persons in the carriage except $A_1, A_2, \ldots, A_{m+1}$. If B is a person other than the $m+1$ persons and he has made friends with at least two of $A_1, A_2, \ldots, A_{m+1}$. Suppose B and A_1, A_2 are friends. Then the m persons $B, A_3, A_4, \ldots, A_{m+1}$ have two common friends A_1, A_2, which contradicts the given condition. So the person B other than $A_1, A_2, \ldots, A_{m+1}$ can make friends with at most one of $A_1, A_2, \ldots, A_{m+1}$. Without loss of generality, we suppose that $A_2, A_3, \ldots, A_{m+1}$ except A_1 cannot be the friends of B. Of course, the common friend C of m passengers $B, A_1, A_2, \ldots, A_{m+1}$ is neither $A_2, A_3, \ldots, A_{m-1}$ nor A_1. Since $m \geq 3$, C makes friends with $m-1 \geq 2$ persons among $A_1, A_2, \ldots, A_{m+1}$. It contradicts the fact that C makes friends with at most one of $A_1, A_2, \ldots, A_{m+1}$. So in the carriage, there are only $m+1$ passengers $A_1, A_2, \ldots, A_{m+1}$ each of whom has m friends.

12 K_5 contains $\binom{5}{2} = 10$ edges and $\binom{5}{3} = 10$ triangles. The number of triangles which is related to every edge is 3. If the number of the edges of the graph is no less than 7, we remove at most 3 edges from K_5. Hence we can remove at most $3 \times 3 = 9$ triangles, so there is still a triangle. It contradicts the condition, so the graph cannot contain 7 or more than 7 edges.

Exercise 2

1 Since $\sum_{i=1}^{n} d(v_i) = 2e$, $n\delta \leq 2e \leq n\Delta$, then $\delta \leq \dfrac{2e}{n} \leq \Delta$.

2 $\sum_{i=1}^{n} d(v_i) = 2(n+1)$, the average degree of the vertices is

$\frac{2(n+1)}{n} > 2$, hence there is at least one vertex whose degree is no less than 3.

3 We denote the faces of a polyhedron by vertices. We join two vertices if and only if the faces which are represented by the vertices have a common edge. We obtain a graph G. By assumption, the number of the odd-degree vertices is odd. So such polyhedron does not exist.

4 No. Use Theorem 2.

5 Construct a graph G as follows. We denote 123 persons by 123 vertices $v_1, v_2, \ldots, v_{123}$. If two persons have discussed the problem, we join the corresponding vertices representing them. So the degree of every vertex in the graph is no less than 5. If the graph contains no vertex whose degree is more than 5, the degree of every vertex in the graph is 5. It means that the number of odd vertices in the graph is odd, which is possible. Then the graph contains at least one vertex whose degree is more than 5.

6 Construct a graph G as follows. We denote n congressmen by n vertices. If two persons do not know each other, we join them in the graph G. For every vertex v_i, $d(v_i) \leq 3$. Now divide G into two components G_1 and G_2. For two vertices in one component, if there is an edge joining them in the original graph, we keep the edge. For two vertices in the different components, if there is an edge joining them in the original graph, the edge will not exist. The removed edges form a set E. In the two components, suppose there is a vertex whose degree is more than 1. Without loss of generality, assume that G_1 contains a vertex v_1, $d(v_1) \geq 2$. Then we move the vertex to G_2. Then G_1 will lose two edges. Since $d(v_1) \leq 3$, G_2 increases by at most one edge and then E increases by at least one edge. Repeat this process. The number of E is increasing, but the total number of edges is finite. So at one point, in the two components, there is no vertex whose degree is more than 1. We complete the proof.

7 The problem can be rephrased as follows. In a graph G with $2n$ vertices, the degree of every vertex is no less than n. Prove that G contains a rectangle. If $G = K_{2n}$, the conclusion is true. If $G \neq K_{2n}$, there exist vertices v_1 and v_2 which are not adjacent. Since $d(v_1) + d(v_2) \geq 2n$, by the Pigeonhole Principle, among the remaining $2n - 2$ vertices there must be two vertices v_3, v_4 which are adjacent to both v_1, v_2. Then the 4 vertices form one rectangle.

8 Construct a graph G as follows. We denote 9 persons by 9 vertices. The two vertices are adjacent if and only if they have shaken their hands. Since $d(v_9) = 6$, there exists $v_k \neq v_1, v_2, v_3$ which is adjacent to v_9. Clearly $d(v_k) \geq 5$. Among the other 5 vertices adjacent to v_9 there is a vertex v_h adjacent to v_k. (Otherwise, $d(v_k) \leq 9 - 5 - 1 = 3$.) Then v_9, v_h, v_k are what we need.

9 We denote the 14 persons by 14 vertices $\{v_1, v_2, \ldots, v_{14}\}$. Two vertices v_i, v_j are adjacent if and only if they have not cooperated, then we get graph G. The degree of every vertex in G is 8. After playing three sets, we will remove 6 edges. So there exist at least two vertices whose degrees are also 8. Let one of them be v_1, among the 8 vertices adjacent to v_1 there must be one vertex whose degree is no less than 7. We know v_2 is adjacent to one vertex v_3 which is another vertex adjacent to v_1. Then v_1, v_2, v_3 and a new vertex v form K_4.

10 Every two vertices are adjacent if the distance between them is given. We prove this problem by induction. When $n = 4$, $\frac{1}{2}n(n-1) + 4 = 6$. Among the 4 vertices, there are 6 distances which are all given. The conclusion is true. Suppose $n = k$ ($k \geq 4$), the conclusion is true. When $n = k + 1$, there are $\frac{1}{2}(k+1)(k-2) + 4$ edges in the vertex set. Let A_{k+1} be the vertex with the smallest degree in the vertex set. Its degree is

$$d(A_{k+1}) \leqslant \frac{2\left[\frac{1}{2}(k+1)(k-2)+4\right]}{k+1}$$

$$= k - 2 + \frac{8}{k+1}$$

$$\leqslant k - 2 + \frac{8}{5} < k.$$

So $d(A_{k+1}) \leqslant k - 1$. Then among the remaining k vertices A_1, A_2, A_3, ..., A_k there are at least

$$\frac{1}{2}(k+1)(k-2) + 4 - (k+1) = \frac{1}{2}k(k-3) + 4$$

edges. By induction, the set of k vertices is stable. Also,

$$d(A_{k+1}) \geqslant \frac{1}{2}(k+1)(k-2) + 4 - \binom{k}{2} = 3,$$

so A_{k+1} is adjacent to at least three vertices among A_1, A_2, ..., A_k. Suppose that A_{k+1} is adjacent to vertices A_1, A_2, A_3 and that $A_{k+1}A_1 = x$, $A_{k+1}A_2 = y$, $A_{k+1}A_3 = z$. It is easy to prove A_{k+1} can be uniquely determined. If not, let A'_{k+1} be another vertex. Also $A'_{k+1}A_1 = x$, $A'_{k+1}A_2 = y$, $A'_{k+1}A_3 = z$, then A_1, A_2, A_3 are all on the perpendicular bisector of $A_{k+1}A'_{k+1}$. It contradicts the hypothesis that there are no three vertices on a common line. Then $A_{k+1}A_4$, ..., $A_{k+1}A_k$ can be determined. The set $\{A_1, A_2, ..., A_{k+1}\}$ is stable. The conclusion is true when $n = k + 1$. In summary, the conclusion is true.

11 We use the unit cube as the vertex. We join the two corresponding vertices if and only if there is a common face between the two unit cubes and we get a graph G. The number of edges of its complementary graph $\overline{G}$ is what we need. It is easy to know the number of edges of G is $3n^2(n-1)$, the number of edges of K_{n^3} is $\frac{1}{2}n^3(n^3 - 1)$ and the number of edges of $\overline{G}$ is

$$\frac{1}{2}n^3(n^3 - 1) - 3n^2(n - 1) = \frac{1}{2}n^6 - \frac{7}{2}n^3 + 3n^2.$$

118 Graph Theory

The number of the pairs of unit cubes which have no more than 2 common vertices is $\frac{1}{2}n^6 - \frac{7}{2}n^3 + 3n^2$ at all.

12 We consider the connected part of the routes which contain the capital. It is a connected graph. What we must prove is that this connected graph contains city A. If not, one vertex (capital) of the graph is adjacent to 21 edges, so every other vertex is all adjacent to 20 vertices. It means that the graph contains only one odd vertex, which is impossible.

Exercise 3

1 Since $\delta |X| = \delta |Y|$ is equal to the total number of edges, then $|X| = |Y|$.

2 We will prove the cases when n is odd and when n is even. Here we only prove the latter. Suppose the theorem is true for all even numbers $n \leqslant 2k$. Let G be a graph which contains $2k + 2$ vertices and no triangle. Remove two adjacent vertices v and v' from G to get G'. By induction, G' contains at most $\left[\frac{4k^2}{4}\right] = k^2$ edges. Since G contains no triangle and any vertex v'' cannot be adjacent to v, v' at the same time, G contains at most

$$k^2 + l + (2k - l) + 1 = k^2 + 2k + 1 = \left[\frac{(2k+2)^2}{4}\right]$$

edges, where l is the number of the vertices adjacent to v in G'.

3 Construct a complete bigraph $K_{10,10}$.

4 Use Theorem 1.

5 (1) Suppose that $n = mk + r$ ($0 \leqslant r < m$). Then by the definition of $T_{m,n}$, $e_m(n) = \binom{n}{2} - r\binom{k+1}{2} - (m-r)\binom{k}{2}$, replace r by $n - mk$ and simplify it to get the solution.

(2) Let the number of the vertices of m parts of a complete m-partite graph G be $n_1, n_2, \ldots, n_m$ respectively. If G is not

isomorphic to $T_m(n)$, then there exists $n_i - n_j > 1$. Consider a complete m-partite graph G', the number of vertices of its m parts is n_1, $n_2, \ldots, n_i - 1, \ldots, n_j + 1, \ldots, n_m$. Since

$$e(G) = \frac{1}{2} \sum_{k=1}^{m} (n - n_k) n_k,$$

$$\begin{aligned} e(G') &= \frac{1}{2} \sum_{k=1, k \neq i, j}^{m} (n - n_k) n_k + \frac{1}{2}(n - n_i + 1)(n_i - 1) \\ &\quad + \frac{1}{2}(n - n_j - 1)(n_j + 1) \\ &= e(G) + (n_i - n_j) - 1 > e(G). \end{aligned}$$

If G' is isomorphic to $T_m(n)$, we complete the proof. Otherwise, repeat the above step until we find a graph which is isomorphic to $T_{m,n}$.

6 Construct a bigraph $G = (X, Y; E)$ as follows. We denote every student from country X by a vertex in X and every student from country Y by a vertex in Y. If a student from X has danced with a student from Y, then join the vertices corresponding to them. Suppose that the degree of x is the largest in X. Since $d(x) < n$, in Y, there is a vertex y' which is not adjacent to x. Suppose that x' in X is adjacent to y'. Since except y', there are $d(x') - 1$ vertices adjacent to x' and $d(x') - 1 \leqslant d(x) - 1 < d(x)$, so there must be a vertex y which is adjacent to x but not adjacent to x'. Then we get four vertices x, x', y, y' corresponding to four persons who satisfy the requirement.

7 Construct a graph G as follows. We denote 14 persons by 14 vertices. Two vertices are adjacent if and only if these two corresponding persons have not cooperated. By assumption, there are $\binom{14}{2} = 91$ pairs among 14 persons. Everyone has teamed up with the other 5 persons, so there are $\dfrac{14 \times 5}{2} = 35$ pairs. Now they play 3 sets and have 6 new pairs. The number of edges of G is $91 - 35 - 6 = 50$, but $e_2(14) = 49$. By Turán's theorem, G contains K_3 and the travelers corresponding to three vertices can play a set with the new traveler.

8 Set $G = (V, E)$. There are $d(x)\{n - 1 - d(x)\}$ triple group $\{x, y, z\}$. They do not form a triangle in G or $\overline{G}$, and $x \in V$ is the end of the only edge in G. Every triple group $\{x, y, z\}$ which does not form a triangle in G or $\overline{G}$ contains one or two edges of G. Suppose that (x, y) is one edge of G and (x, z), (y, z) are two edges of $\overline{G}$. In the sum $\sum_{x \in V} d(x)\{n - 1 - d(x)\}$, the triple group $\{x, y, z\}$ has been counted twice: one is about x, the other is about y. If (x, y), (y, z) are the edge of G, (x, z) is the edge of $\overline{G}$, in above sum, the tuple group $\{x, y, z\}$ has also been counted twice: one involves x, the other involves z. The sum of triangles in G and $\overline{G}$ is

$$\binom{n}{3} - \frac{1}{2} \sum_{x \in V} d(x)(n - 1 - d(x))$$

$$\geq \binom{n}{3} - \frac{n}{2}\left(\frac{n-1}{2}\right)^2$$

$$= \frac{1}{24} n(n-1)(n-5).$$

9 Suppose there is no $(k + 1)$-element set which we need. We will prove

$$m \leq \frac{(k-1)(n-k) + k}{k^2} \binom{n}{k-1}.$$

We denote all the red k-subsets by S and all the $(k - 1)$-subset by β. For any $(k - 1)$-subset B, we denote the number of red k-element subsets which contain B by $\alpha(B)$. For any $A \in S$, A contains $k(k - 1)$-element subsets. For any element $x \in X \backslash A$, x together with at most $k - 1$ of $k(k - 1)$-element subsets form a red k-element subset. (Otherwise, there exists $(k + 1)$-element subset and all its k-element subsets are red k-element subsets.) So

$$\sum_{B \subset A, |B| = k-1} \alpha(B) \leq (n-k)(k-1) + k.$$

So

$$m[(n-k)(k-1)+k] \geqslant \sum_{A \in SB \subset A, |B|=k-1} \sum \alpha(B)$$
$$= \sum_{B \in \beta}(\alpha(B))^2$$
$$\geqslant \frac{1}{|\beta|}\left(\sum_{B \subset \beta}\alpha(B)\right)^2$$
$$\geqslant \frac{1}{\binom{n}{k-1}}(km)^2.$$

So

$$m \leqslant \frac{[(n-k)(k-1)+k]\binom{n}{k-1}}{k^2}.$$

Since

$$m > \frac{[(n-k)(k-1)+k]\binom{n}{k-1}}{k^2},$$

there must exist a $(k+1)$-element subset of X such that all k-element subsets are red k-element subsets.

10 Since $\binom{10}{2} = 45$, then a complete graph with 10 vertices contains 45 edges. The figure in the problem is obtained by removing 5 edges from the complete graph with 10 vertices. We call the 5 edges "Removed Edges" and denote 10 vertices by A_1, A_2, ..., A_{10}. Without loss of generality, let A_1A_2 be a "Removed Edge", then remove A_1 and its incident edges. The deduced graph with 9 vertices contains at most 4 "Removed Edges". Without loss of generality, let A_2A_3 be a "Removed Edge". Then remove A_2 and its incident edges. (If there is no "Removed Edge", remove any vertex. The same in the later step.) The deduced graph with 8 vertices contains at most 3 "Removed Edges". Without loss of generality, let A_3A_4 be a "Removed Edge". Then remove A_3 and its incident edges. The deduced graph with 7 vertices contains at most 2 "Removed Edges".

Without loss of generality, let A_4A_5 be a "Removed Edge". Then remove A_4 and its incident edges. The deduced graph with 6 vertices contains at most 1 "Removed Edge". The deduced graph is a complete graph with 6 vertices or the graph which is obtained by removing one edge from the complete graph with 6 vertices. Anyway, the graph must contain a complete bipartite graph $K_{3,3}$. We can generalize this problem: If a graph with n vertices and m edges contains no $K_{n,n}$. We can prove that $m < C \cdot n^{2-\frac{1}{r}}$, where C depends on r.

11 We denote the positions of 18 police cars by 18 vertices $x_1, x_2, \ldots, x_{18}$. Suppose

$$E = \left\{(x_i, x_j) \,\bigg|\, \frac{d(x_i, x_j)}{12} \leqslant \frac{\sqrt{2}}{2} < \frac{9}{12}\right\}.$$

By Theorem 3, $|E| \geqslant \binom{18}{2} - \left\lceil \frac{18^2}{3} \right\rceil = 45$. It means there are at least 45 pairs of cars which can communicate with each other. If the above condition of the graph does not hold, then there does not exist two vertices whose degrees are more than 5 and

$$|E| \leqslant \frac{1}{2}(1 \times 17 + 4 \times 17) < 43,$$

a contradiction.

12 When $n = 2$, $n^2 + 1 = 5$. There are 5 line segments among 4 vertices, which form two triangles. Suppose that proposition is true for $n = k$. When $n = k + 1$, let us prove that there exists at least one triangle. Suppose that AB is a given line segment and denote the number of line segments from A and B to other $2k$ points by a and b.

(1) If $a + b \geqslant 2k + 1$, there exists a vertex C other than A and B so that AC and BC exist. Then there exists a triangle $\triangle ABC$.

(2) If $a + b \leqslant 2k$, if we remove A and B, among the remaining $2k$ vertices, there exist at least $k^2 + 1$ line segments. By induction, there exists a triangle.

Suppose $\triangle ABC$ is a triangle formed by these line segments. We denote the number of line segments from A, B and C to other $2k-1$ points by α, β and γ.

(3) If $\alpha+\beta+\gamma\geqslant 3k-1$, the total number of the triangles including one of AB, BC, CA as an edge is at least k. These k triangles together with $\triangle ABC$ give $k+1$ triangles.

(4) If $\alpha+\beta+\gamma\leqslant 3k-2$, there is at least one number which is no more than $2k-2$ among the three numbers $\alpha+\beta$, $\beta+\gamma$, $\gamma+\alpha$. Without loss of generality, $\alpha+\beta\leqslant 2k-2$. When we remove two vertices A, B, among the remaining $2k$ vertices there exist at least k^2+1 line segments. By induction, there exist at least k triangles. These k triangles together with $\triangle ABC$ give $k+1$ triangles. The proposition is true when $n=k+1$. We complete the proof by induction.

Exercise 4

1 The spanning tree of graph G contains two pendant vertices. Remove these two vertices and the graph is still connected.

2 There are $9\times 9=81$ vertices whose degrees are 4, so we should remove at least $\left[\dfrac{81}{2}\right]+1=41$ edges so that the degree of every vertex is less than 4. We can remove at most $2\times 11\times 10-120=100$ edges so that the graph is still connected.

3 The proposition is not true. Take K_3 and an isolated vertex (the vertex which is not adjacent to any other vertex) to construct a graph G. Then G contains 4 vertices and 3 edges, so it is not connected and clearly not a tree.

4 (1) Suppose that T has x pendant vertices. The number of vertices of the tree T is $n=3+1+x$, the edge number is $e=n-1=x+3$. $\sum\limits_{i=1}^{n}d(v_i)=3\times 3+2\times 1+1\times x=11+x$, so $11+x=2(x+3)$, $x=5$.

(2) Fig. 4 shows us two trees satisfying the requirement but they are not isomorphic.

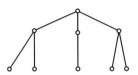

Fig. 4

5 Suppose T contains n vertices and e edges. Then $n = \sum_{i=1}^{k} n_i$, $e = n - 1$,

$$\sum_{i=1}^{n} d(v_i) = \sum_{i=1}^{k} i n_i = 2e = 2n - 2 = 2\sum_{i=1}^{k} n_i - 2.$$

So

$$n_1 = \sum_{i=2}^{k} (i-2)n_i + 2.$$

For $r \geq 3$, by the above equality, we can obtain

$$n_r = \frac{1}{r-2}\Big[\sum_{\substack{i=1\\i\neq r}}^{k} (2-i)n_i - 2\Big].$$

6 Among $d_1, d_2, \ldots, d_n$, there must be at least two which is equal to 1. $\Big($Otherwise, $\sum_{i=1}^{n} d_i \geq 2n - 1\Big)$. We apply induction on the number of vertices n. When $n = 2$, the proposition is true. Suppose that the conclusion is true when $n = k$. When $n = k + 1$, there exists a number 1 among $d_1, d_2, \ldots, d_k, d_{k+1}$. Without loss of generality, let $d_{k+1} = 1$. It is easy to know among the $k + 1$ numbers there exists a number which is no less than 2, denoted by d_k. Consider the k numbers $d_1, d_2, \ldots, d_{k-1}, (d_k - 1)$,

$$d_1 + \cdots + d_{k-1} + (d_k - 1) = 2(k+1) - 2 - 1 - 1 = 2k - 2.$$

By induction, there exists tree T' whose vertices are $v_1, \ldots, v_k$,

$$\sum_{i=1}^{k} d(v_i) = d_1 + \cdots + d_{k-1} + (d_k - 1) = 2k - 2.$$

In T', there is an edge which is from v_k to v_{k+1}. We obtain a tree

T, then

$$\sum_{i=1}^{k+1} d(v_i) = 2k - 2 + 1 + 1 = 2(k+1) - 2.$$

T is what we need.

7 Construct a graph G, we use the ends of n line segments as the vertices of G and the line segments as the edges. By the condition, G is connected and contains no loop. So G is a tree where the length of its longest chain is 2. So G contains only one vertex which is not a pendant vertex. The vertex is the common vertex of n line segments.

8 Refer to Example 6 in this chapter.

9 Suppose the conclusion is not true, then there must exist a counterexample. Consider the counterexample in which $|E|+|V|$ is the smallest. In this counterexample, $|E|=|V|+4$. (Otherwise, we can remove more edges and still get a counterexample, where $|E|+|V|$ is smaller. A contradiction!) Then $|E|>|V|$. The graph must contain a cycle. The length of the shortest cycle is at least 5. (Otherwise, the length of the shortest cycle is no more than 4, then we remove this cycle. We still have $|E|\geqslant|V|$. There still exists a cycle. The cycle and the above cycle contain a common edge. A contradiction!)

Furthermore, the degree of every vertex is at least 3. Otherwise, if the degree of a vertex is 2, remove this vertex and change the two edges adjacent to this vertex to one edge. We still have $|E|=|V|+4$ and $|E|+|V|$ gets smaller. A contradiction! If the degree of one vertex is 1, remove this vertex and its adjacent edges, we still have $|E|=|V|+4$ and $|E|+|V|$ is smaller. A contradiction! If there exists an isolated vertex, remove it and $|E|>|V|+4$ and $|E|+|V|$ is smaller. A contradiction! Take a cycle C_0, the length of which is at least 5. The cycle contains at least 5 vertices. For every vertex on the C_0, it is adjacent to at least one edge which is not on the cycle and the adjacent vertices are distinct. (Otherwise, there exists a cycle whose length is less than 5.) Then it is easy to see $|V|\geqslant 2\times 5 = 10$.

On the other hand, $2\mid E\mid = \sum_{v\in V}d(V) \geqslant \sum_{v\in V}3 = 3\mid V\mid$. $\mid E\mid = \mid V\mid + 4$, so $2\mid V\mid + 8 \geqslant 3\mid V\mid$, $\mid V\mid \leqslant 8$. A contradiction! Such counterexample does not exist. We have proven the proposition.

Remark The method of proof in this problem is the proof by contradiction. To prove that the proposition is true, we assume that the proposition is false. Consider a variable $V \in$ **N**. From the smallest counterexample of V we can deduce a contradiction and the proof becomes easier using this condition. The conclusion of this problem is the best. When $\mid E\mid = \mid V\mid + 3$, we can give a counterexample as Fig. 5 shows us.

Fig. 5

10 We denote 21 persons by 21 vertices. There is an edge joining two vertices if and only if the two delegates which are represented by the two vertices have called each other. By assumption, there exists an odd cycle whose length is m. (We call cycle whose length is odd cycle.)

Let C be the smallest odd cycle in the graph, the length of which is $2k+1$.

If $k=1$, let C be a triangle. It means the three persons have made a phone call to each other.

If $k>1$, set C $v_1v_2\ldots v_{2k+1}v_1$ and there is no edge joining v_i and v_j. ($1\leqslant i,j\leqslant 2k+1$, $i-j\neq \pm 1 \pmod{2k+1}$.) Otherwise, suppose that v_i, v_j are adjacent and the sum of the length of cycle $v_1v_2\ldots v_iv_j\ldots v_{2k+1}v_1$ and that of the cycle $v_iv_{i+1}\ldots v_jv_i$ is $2k+3$. So among them there must be an odd cycle whose length is less than $2k+1$. It contradicts the fact that C is the shortest.

Suppose there is no triangle among the $21-(2k+1)=20-2k$ vertices other than $v_1, v_2, \ldots, v_{2k+1}$. By Turán's theorem, there are at least $(10-k)^2$ edges joining them. Any vertex among them is not adjacent to two vertices which are adjacent to C, so it is adjacent to at most k vertices. So the sum of the edges is:

$$2k + 1 + k(20 - 2k) + (10 - k)^2$$
$$= 100 + 2k + 1 - k^2$$
$$= 102 - (k - 1)^2$$
$$\leqslant 102 - (2 - 1)^2 = 101.$$

A contradiction!

So the graph must contain a triangle, which means there exist three persons who have called each other.

11 Suppose there exists such a graph that the degree of every vertex is more than 2. But the length of any cycle of the graph is divisible by 3. We consider the graph G which has this property and the least number of vertices. Clearly, the graph contains the shortest circle Z. The non-adjacent vertices on this cycle are not joined by an edge. Since the degree of every vertex is more than 2, every vertex on the cycle Z is adjacent to one vertex not on the cycle. Let Z pass the vertex $A_1, A_2, \ldots, A_{3k}$.

Suppose that there exists a path S which joins the vertices A_m and A_n and which does not include edges in Z. We consider the cycle Z_1 and Z_2 consisting of the two halves of S and Z. Since the length of each of the two cycle is divisible by 3, it is not difficult to see the length of path S is divisible by 3. Especially, for the given graph, we can know that any vertex X which is not on the circle Z cannot have edges incident to two distinct vertices of Z. It means that the edges which are induced by the vertices on Z but not on the cycle should be incident to distinct vertices, respectively.

Let us construct a graph G_1. Collapse all vertices $A_1, A_2, \ldots, A_{3k}$ on the cycle Z of G into one vertex A and keep all the vertices which are not on the cycle and their incident edges. Join the A and the vertices on the Z one by one. It is easy to know the degree of A is no less than $3k$. The number of vertices in G_1 is less than that of G and the degree of every vertex is still more than 2. According to above conclusion, the length of any cycle in G is divisible by 3. We arrive at a contradiction. In view of the above proof, we can know G is the graph satisfying these properties and with the least vertices.

Then in a graph the degrees of its vertices are all more than 2, there must exist a cycle whose length can be divided by 3. Then we only need to apply this assertion to our problem. We denote the city by a vertex and the path by the edge.

Exercise 5

1 When n ($n \geq 2$) is odd, K_n is a cycle. When $n = 2$, K_2 is a chain. When m, n are both even, $K_{m,n}$ is a cycle.

2 Suppose G contains at least $2k$ odd vertices. Delete one edge to get G'. There are three cases: (1) The number of odd vertices of G' decreases by 2, then G' need at least $k - 1$ strokes to draw. (2) The number of odd vertices increases by 2, G' need at least $k + 1$ strokes to draw. (3) The number of odd vertices does not change, G' need at least k strokes to draw.

3 These two graphs are all unicursal, that is, they can be drawn in one strock, and start and end at the same vertex.

4 When n is odd, the graph is unicursal; when n is even, the graph is not unicursal.

5 Draw G as follows. We denote the persons by vertices. If two person have exchanged views, then join the corresponding vertices. Take the longest chain μ. Let v_1 be an end of μ, then δ vertices $v_2, \ldots, v_{\delta+1}$ adjacent to v_1 are all on the chain μ. Otherwise, μ can still be extended. Go along the μ through vertices $v_2, v_3, \ldots, v_{\delta+1}$, and then return to v_1. This is a cycle whose length is more than δ.

6 Take a point v_j' ($j = 1, 2, 3, 4$) on every face. If the two faces have common edges, join the two points. Then we get a new graph G^* which we call the dual graph of G. In the graph G, going from a face to another face through the edge e_i is equivalent to going from one vertex to another vertex along an edge in the G^*. Therefore, if G

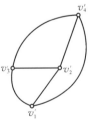

Fig. 6

contains a broken line μ satisfying conditions (1) and (2), then G^* is a chain (Q_1 and Q_2 are not on a face) or a cycle (Q_1 and Q_2 are on a face), i.e. the graph G^* is unicursal or it can be drawn in one strock. But if the four vertices of G^* are all odd, the graph G^* needs two strocks to draw.

7 Suppose that there are k lines and that one vertex v_i corresponds to one number a_i in the following way. If v_i is red, then $a_i = 1$; v_i is blue, then $a_i = -1$, $i = 1, 2, \ldots, n$. Then

$$-1 = a_1 a_n = (a_1 a_2)(a_2 a_3) \ldots (a_{n-1} a_n) = (-1)^k,$$

hence k is odd.

8 Use the conclusion of Exercise 7 and refer to Example 5 in Chapter 5.

9 The given graph contains 16 odd vertices B_i, C_i ($i = 1, 2, \ldots, 8$). If we want to make it a cycle, we should add at least 8 edges so that the graph becomes a cycle. Fig. 7 shows the cycle after adding 8 edges $B_i C_i$ ($i = 1, 2, \ldots, 8$) so that the walk is the shortest.

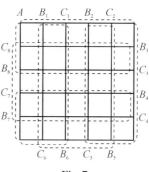

Fig. 7

Exercise 6

1 When $n \geqslant 3$, K_n is a Hamiltonian graph. When $m = n \geqslant 2$, the complete bigraph $K_{m,n}$ is a Hamiltonian graph.

2 The reader can find it on the graph.

3 A regular icosahedron consists of 20 congruent equilateral triangles. At the center of every triangle we mark a vertex. Only if two triangles have a common edge, we join the corresponding vertices and construct a regular dodecahedron which consists of 12 regular pentagons. From the study of Hamiltonian cycles, we can know that on the regular dodecahedron we can find a Hamiltonian cycle. Use scissors to cut the dodecahedron along the Hamiltonian cycle. It means that cut the regular dodecahedron into two halves, and also cut each regular triangle into two halves. The trace does not go through the vertices of the regular dodecahedron.

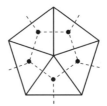

Fig. 8

4 Draw a graph G as follows. We denote each $1 \times 1 \times 1$ cube by a vertex. We join the corresponding vertices if and only if two cubes contain a common face. It is easy to know that G is a bigraph. Let $G = (X, Y; E)$. If the corresponding vertex corresponding to the small cube at one corner belongs to X, the vertex which represents the small cube at the center belongs to Y. Since $|X| = 14$, $|Y| = 13$, G contains no Hamiltonian chain.

5 (1) We denote six persons by six vertices $v_1, v_2, \ldots, v_6$. If v_i can cooperate with v_j, v_i is adjacent to v_j. By condition, $d(v_i) \geq 3$, $i = 1, 2, \ldots, 6$. According to Theorem 4, G contains a Hamiltonian cycle $C = v_{i_1} v_{i_2} \ldots v_{i_3} v_{i_1}$. In the cycle, two adjacent vertices represent two persons who can cooperate with each other.

(2) Put v_{i_1}, v_{i_2} into one group, v_{i_3}, v_{i_4} into one group, v_{i_5}, v_{i_6} into one group. Alternatively put v_{i_6}, v_{i_1} into one group, v_{i_2}, v_{i_3} into one group, v_{i_4}, v_{i_5} into one group. These are two different ways of grouping.

6 We denote $2n$ ministers by $2n$ vertices. If two persons are not enemies, join the corresponding vertices to get a graph G. In the graph G, for every vertex v, $d(v) \geq (2n - 1) - (n - 1) = n$. According to the Theorem 4, G contains a Hamiltonian cycle and we can arrange vertices according to the order in the cycle.

7 Draw a graph G as follows. We denote 9 children by 9 vertices and if two children know each other we join the corresponding vertices. In graph G, for any two vertices v and v', $d(v) + d(v') \geq 8$. According to Theorem 2, G contains a Hamiltonian chain and we can arrange the children into one line by the order in the chain.

8 Draw a graph G as follows. We denote the materials by vertices and every dish by an edge. In graph G, the degree of every vertex is no less than 4. According to Theorem 4, G contains a Hamiltonian cycle.

9 Suppose that set A contains n elements and we give every element a number. Let $A = \{1, 2, 3, \ldots, n\}$. We use a sequence whose length is n and consisting of 0 and 1 to denote a subset. The rule is that if the element i of A is in this subset, the i-th position of this sequence is 1, otherwise the i-th position is 0. For example, the empty set $\varnothing = 0, 0, 0, \ldots, 0$; $\{1\} = 1, 0, 0, \ldots, 0$; $\{n\} = 0, 0, \ldots, 1$; $\{2, 3\} = 0, 1, 1, 0, \ldots, 0$. Then there are 2^n subsets in A. We denote the sequences corresponding to the 2^n subsets by vertices and join two vertices if and only if the sequences have only one different number at the same position. Then we get a graph G. For example, when $n = 1$, G is a line segment as Fig. 9 shows us; When $n = 2$, G is a square as Fig. 10 shows us. Fig. 10 can be drawn using two Fig. 9. That is, add 0 before one pair of 0, 1 to get 00, 01 and add 1 before another pair of 0, 1 to get 10, 11. Then put one Fig. 9 on top of another one and join them by two vertical edges to get a square. Copy two Fig. 10, put one on top of another, add 0 before the number of every vertex of the top square and add 1 before the number of every vertex of the bottom square. Join the corresponding vertices of the two squares by four vertical edges to get a graph G. When $n = 3$, G is a cube. If $n = k$, assume that the graph G has been drawn. Put G on top of another copy. Then add 0 before the number of every vertex of the upper graph and add 1 before the number of every vertex of below graph. Join the corresponding vertices of the two graphs by vertical edges to get a new graph G of $n = k + 1$. Fig. 11 is the case when $n = 3$ and Fig. 12 is the case of $n = 4$. We call the graph when $n = k$ a cube graph with dimensions n. It is easy to prove by induction: the cube graph with dimensions n contains a Hamiltonian cycle ($n \geq 2$). For $n = 1$, clearly it is true. Since the cube graph with dimension 1 is K_2 which is a Hamiltonian chain. For $n = 2$, it is a square which is a Hamiltonian cycle. For $n = k$, if it is a Hamiltonian graph, consider $n = k + 1$. Delete one corresponding edge from the upper and lower Hamiltonian cycle of $n = k$ in G, respectively and then combine the edges joining the ends of deleted edges and the upper and lower Hamiltonian cycle to get a Hamiltonian cycle when $n = k + 1$ as the bold lines of Fig. 12 show us. Put the vertices of G on a cycle according to the order of Hamiltonian differs only cycle. Start from any vertex and order all subsets clockwise (or counter clockwise) so that every adjacent subset differs only in one element.

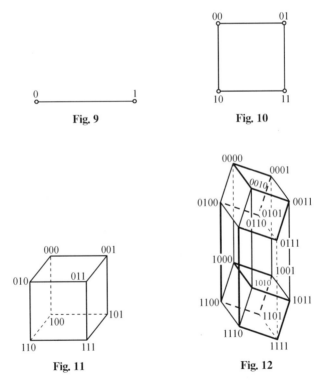

Fig. 9   Fig. 10

Fig. 11   Fig. 12

10 First, the degree of every vertex is at least 3. Otherwise, there exists one vertex A which is incident to at most two edges.

Remove one edge, then the remaining vertex A cannot lie on cycle which contradicts the condition. So $n \geq 3$. It is easy to prove that $n \neq 4, 5, 6$.

If $n = 7$, remove the vertex whose degree is the largest (clearly, the degree is at least 3) to get a cycle whose length is 6. Since the vertices adjacent to this vertex may be non-adjacent on the cycle. (Otherwise, there will be a cycle whose length is 7.) The removed vertex is adjacent to at most three non-adjacent vertices on the cycle. So the degree of this vertex is at most 3. $3 \times 7 = 21$ is odd. In fact, the sum of all the degrees is even. A contradiction.

If $n = 8$, after removing the vertex whose degree is the largest, we get a cycle whose length is 7. The removed vertex is adjacent to at most three non-adjacent vertices on the cycle. So the degree of this vertex is at most 3. The degree of every vertex is 3. As Fig. 13(1) shows us, the degrees of A, C, F, O are 3. They cannot be incident to any edge. Every vertex of B, D, E, G is incident to one edge, respectively. If B is adjacent to G, D is adjacent to E (there are two edges). It is impossible. If B is adjacent to D, E is adjacent to G and the graph contains a cycle whose length is 8. A contradiction. If B is adjacent to E, D is adjacent to G and the graph contains a cycle whose length is also 8. A contradiction.

If $n = 9$, since $3 \times 9 = 27$ is not even, it is impossible that the degree of every vertex is 3. There exists one vertex whose degree is at least 4. We remove the vertex whose degree is the largest to get a cycle whose length is 8. So the removed vertex is adjacent to at most four non-adjacent vertices on the cycle. The largest degree is 4 and the smallest is 3. As Fig. 13(2) shows us, B is at least incident to one edge. Clearly we cannot join more edges between B and A, C. If B is adjacent to D, the graph contains a cycle whose length is 9. A contradiction. Similarly, B cannot be adjacent to H. If B is adjacent to F, the graph contains a cycle whose length is also 9. A contradiction. So B can only be adjacent to E or G. By symmetry, let B be adjacent to E. Similar to the above argument, we can know H is adjacent to C. (If H is adjacent to E, the degree of E is 5. A contradiction.) F is adjacent to A, and D is adjacent to G. We cannot join any two vertices by more edges and after removing A, the graph must contain a cycle whose length is 8. In fact, if the graph contains a cycle whose length is 8, BE, BC, HG, HC, FE, FG must lie on the cycle. The six edges form a cycle whose length is 6. A contradiction. By the above discussion, the n satisfying the condition is at least 10. The example when $n = 10$ is Fig. 13(3), we call it the Peterson graph.

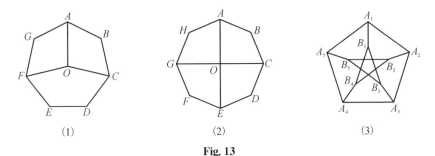

Fig. 13

11 If there are 5 persons, let the original order of seats be $ABCDEA$. Change it to $ADBECA$. If there are more than five persons, join the two vertices only when they are not seated together. Get a graph G. Since the degree of every vertex is $|V(G)|-3$. The sum of the degree of any two vertices is $2n-6$ where n is the number of vertices. Also $n > 5$, $2n-6 \geqslant n$. By Theorem 3, G contains a Hamiltonian cycle and we can arrange the seats by the order of the cycle.

Exercise 7

1 Suppose that G is connected, otherwise consider one connected component. If the degree of every vertex is no less than 6, $6v \leqslant 2e$. It means that $v \leqslant \dfrac{e}{3}$. Since $f \leqslant \dfrac{2e}{3}$, then

$$2 = v - e + f \leqslant \dfrac{e}{3} - e + \dfrac{2e}{3} = 0.$$

A contradiction.

2 Suppose the degree of every vertex is more than 4, then $2e = \sum_{i=1}^{v} d(v_i) \geq 5v$, i.e. $v \leq \frac{2}{5}e$. Since $e \leq 3v - 6$, then $e \leq \frac{6}{5}e - 6$. It means that $e \geq 30$. A contradiction.

3 By Euler's Formula, $f = 2 + e - v = 8$. Since there are $\frac{2e}{f} = 3$ edges on every face in average. Since there are at least three edges on every face, there are 3 edges on every face.

4 Suppose G and $\overline{G}$ are both planar graphs. The number of vertices in G and $\overline{G}$ is v and the number of edges is e and e', respectively. Then $e + e' = \frac{1}{2}v(v-1)$. Adding the inequalities $e \leq 3v - 6$, and $e' \leq 3v - 6$, we get $\frac{1}{2}v(v-1) = e + e' \leq 6v - 12$, $v^2 - 13v + 24 \leq 0$, $v \leq 11$. It contradicts the hypothesis.

5 Consider the dual graph. (See the sixth problem of Exercise 5.) Since K_5 is not a planar graph, then $f \leq 4$.

6 There are $\binom{n}{2}$ edges in a convex polyhedron with n vertices and every face contains at least 3 edges. So the number of the faces of the polyhedron is no more than $\frac{2}{3}\binom{n}{2}$. By Euler's Formula,

$$n + \frac{2}{3}\binom{n}{2} \geq \frac{2}{3}\binom{n}{2} + 2.$$

Simplify it to get $n^2 - 7n + 12 \leq 0$, where n can take only 3 or 4. We complete the proof.

7 See Question 1 of Exercise 7.

8 We denote the vertices of a polyhedron by the vertices of a graph and the edges by the edges of the graph. Then we get a connected planar graph. Then $v \geq 4$, $f \geq 4$. By Euler's Formula $e = v + f - 2 \geq 6$. It means that there is no polyhedron whose edge number is less than 6. If there is a graph with $e = 7$, then $3f \leq 2 \times 7$,

$f = 4$. But a polyhedron with four faces can only contain 6 edges. So there is no polyhedron with 7 edges. Consider $k \geq 4$. The pyramid whose base is a polygon with k edges is a polyhedron with $2k$ edges. Cut a corner from the pyramid whose base is a polygon with $k-1$ edges to get a polyhedron with $2k+1$ edges. In conclusion, $n \geq 6$, $n \neq 7$, there is a polyhedron with n edges.

9 Suppose a convex polyhedron contains x vertices and the $10n$ faces contain $C_1, C_2, \ldots, C_{10n}$ vertices and $a_1, a_2, \ldots, a_{10n}$ edges, respectively. The number of the edges of the convex polyhedron is $\frac{1}{2}\sum_{i=1}^{10n} a_i$. By Euler's Theorem,

$$10n + x = \frac{1}{2}\sum_{i=1}^{10n} a_i + 2.$$

Since $x \leq \frac{1}{3}\sum_{i=1}^{10n} a_i$, then

$$\frac{1}{2}\sum_{i=1}^{10n} a_i + 2 - 10n \leq \frac{1}{3}\sum_{i=1}^{10n} a_i,$$

or

$$\sum_{i=1}^{10n} a_i \leq 60n - 12.$$

If among $10n$ faces there are no n faces whose edge numbers are equal,

$$\sum_{i=1}^{10n} a_i \geq (3 + 4 + \cdots + 12)(n - 1) + 13 \times 10$$

$$= 75n + 55$$

$$> 60n - 12,$$

a contradiction. So there are at least n faces whose edge numbers are equal.

10 In the graph there are only two-side polygons and hexagons and the number of each kind is 3. If the graph contains a Hamiltonian cycle, by Theorem 4, $4(f_6' - f_6'') = 0$. That is $f_6' = f_6''$, but $f_6' + f_6'' = 3$. It is impossible.

11 By Theorem 4, $2(f_4' - f_4'') + 3(f_5' - f_5'') = 0$, so $f_4' - f_4''$ is a multiple of 3. It means among the 5 quadrilaterals there are 4 quadrilaterals outside the cycle, and another inside, or 4 quadrilaterals inside. If this Hamiltonian cycle goes through both e and e', the two quadrilaterals on the two sides of e are inside the Hamiltonian cycle and outside the cycle, respectively. So are the two quadrilaterals on the two sides of e'. There are at least two quadrilaterals inside the cycle and two others outside. A contradiction.

12 Draw a graph $G = (V, E)$, where $V = \{x_1, x_2, \ldots, x_n\}$. In the graph G, two vertices x_i, x_j are adjacent if and only if $d(x_i, x_j) = 1$. Suppose that G contains two distinct edges AB, CD which intersect at vertex O as Fig. 14 shows us. Since $d(A, B) = 1$, $d(C, D) = 1$, without loss of generality, we suppose that $d(O, A) \leqslant \frac{1}{2}$, $d(O, C) \leqslant \frac{1}{2}$ and the angle between AB and CD is θ, $d(A, C) = \{d^2(O, A) + d^2(O, C) - 2d(O, A) \times d(O, C) \cos \theta\}^{\frac{1}{2}}$. By the above condition only when $\theta = \pi$, $d(O, A) = \frac{1}{2}$, $d(O, C) = \frac{1}{2}$, then $d(A, C) = 1$. But now A coincides with D and B coincides with C, it means that AB and DC are the same edge. It contradicts the hypothesis that they are two distinct edges. Other than this case, we have $d(A, C) < 1$. It contradicts the hypothesis that the distance of any two vertices is no less than 1. In summary, G is a planar graph and the edge number e of G is no more than $3n - 6$. (The reader can compare it with the conclusion of Example 7 in Chapter 2.)

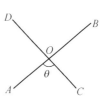

Fig. 14

Exercise 8

1 By Example 1, there must exist a monochromatic triangle.

Without loss of generality, let $\triangle A_1A_2A_3$ be red. For the three edges of $\triangle A_4A_5A_6$, there are two cases: (1) The three edges of $\triangle A_4A_5A_6$ are all red. We have completed the proof. (2) There is one blue edge in $\triangle A_4A_5A_6$, say, A_4A_5. For A_1A_4, A_2A_4, A_3A_4, if there are two red edges among them, there must exist a red triangle. If there are two blue edges, say, A_1A_4 and A_2A_4. Now if there is one blue edge among A_1A_5 and A_2A_5, there must exist a blue triangle. If A_1A_5 and A_2A_5 are all red, $\triangle A_1A_2A_5$ is a red triangle. We complete the proof.

2 If there are no three line segments such that they are monochromatic and contain no common vertices. As Fig. 15 shows us, without loss of generality, let A_1A_2 be red. By hypothesis, three line segments A_3A_4, A_5A_8, A_6A_7 cannot be all blue. Without loss of generality, let A_3A_4 be red. Since A_1A_2 and A_3A_4 are red, the lines joining every two of the four points A_5, A_6, A_7,

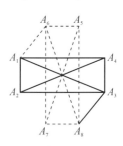

Fig. 15

A_8 are all blue. Similarly, the lines joining every two of the four points A_1, A_2, A_3, A_4 are all red. Without loss of generality, let A_1A_6 be blue, then A_3A_8 must be red. Whatever color A_4A_5 is, it contradicts the hypothesis.

3 We color the shortest edge of every triangle red and others blue. Since $r_2 = 6$, there must be a monochromatic triangle which is red and whose longest edge is the shortest edge of another triangle.

4 Take one vertex A of the two-color complete graph K_9. If A is incident to four blue edges AA_1, AA_2, AA_3, AA_4, the complete subgraph K_4 whose vertices are A_1, A_2, A_3, A_4 in K_9 contains no blue edge. If A is incident to six red edges AA_1, AA_2, ..., AA_6, the complete subgraph whose vertices are A_1, A_2, ..., A_6 in K_9 contains a monochromatic triangle $\triangle A_iA_jA_k$ ($1 \leqslant i, j, k \leqslant 6$). Since K_9 contains no blue triangle, $\triangle A_iA_jA_k$ is a red triangle and the complete subgraph K_4 whose vertices are A, A_i, A_j, A_k is red. If every vertex of K_9 is incident to 5 red edges, then the number of red edges in K_9 is

$\frac{5 \times 9}{2}$. It is impossible.

5 By Theorem 3,
$$r(3, 6) \leqslant r(3, 5) + r(2, 6) - 1 = 14 + 6 - 1 = 19.$$

6 By Theorem 3, $r(4, 4) \leqslant r(4, 3) + r(3, 4) = 9 + 9 = 18$.

7 Consider the n-color complete graph K_{r_n}. The method of coloring is that color (x, y) in the i-th color if and only if $|x - y|$ is in the i-th group. By Theorem 2, K_{r_n} must contain a monochromatic triangle. Suppose that the three edges of the triangle is all colored in the j-th color, then among $1, 2, \ldots, r_n$, there are three natural numbers $a > b > c$ such that $x = a - c$, $y = a - b$, $z = x - y$ are all in the j-th group.

8 By Theorem 2, $r_6 \leqslant [6!e] + 1 = 1958 < 1978$. By Schur's Theorem, the proposition is true.

9 In $A_1, A_2, \ldots, A_7$, among the triangles constructed from the first six vertices there must be two monochromatic triangles which have no common edge. If there is a common edge, let the two triangle be $\triangle A_1 A_2 A_3$ and $\triangle A_1 A_2 A_4$. Now remove A_1 and add A_7, then there exist two monochromatic triangles. Between the two triangles, there must be a triangle different from $A_2 A_3 A_4$. There is no common edge between this triangle and $\triangle A_1 A_2 A_3$ or $\triangle A_1 A_2 A_4$.

10 We denote the six lines by six vertices. If two lines lie on two different faces, color the edge joining the corresponding vertices red. If the lines lie on one face, color the edge joining the corresponding vertices blue. We obtain a two-color complete graph K_6. So there must exist a monochromatic triangle. If it is a red triangle, any two of the lines which the vertices correspond to lie on different faces. If it is a blue triangle, the three lines which the vertices correspond to lie on one face. Since any three lines cannot lie on one face, the three faces which the three lines lie in intersect each other in the three lines. Then we know the three lines either are parallel to each other or intersect at one point.

11 Take a set consisting of four irrational numbers $\{\sqrt{2}, -\sqrt{2}, \sqrt{3}, -\sqrt{3}\}$. Take any three numbers from the set and there must be: either $\sqrt{2} + (-\sqrt{2}) = 0$ is a rational number when $\sqrt{2}, -\sqrt{2}$ are chosen; or $\sqrt{3} + (-\sqrt{3}) = 0$ is a rational number when $\sqrt{3}, -\sqrt{3}$ are chosen. When $n = 4$, the conclusion is not true. So if the n satisfies the requirement of the problem, $n \geq 5$. Next, we prove that among any five given numbers, we may find three such that the sum of any two irrational numbers is still an irrational number. Suppose that $\{x, y, z, u, v\}$ is a set of any given five irrational numbers. We denote the five irrational numbers by five vertices. If the sum of two numbers is an irrational number, join the corresponding vertices by a red edge. If the sum of two numbers is a rational number, join the corresponding vertices by a blue edge. Then we get a two-color complete graph K_5. First we prove that the two-color complete graph K_5 contains no blue triangle. Otherwise, suppose that there is a blue triangle $\triangle xyz$ which means that $x + y, y + z, z + x$ are all rational numbers. Then

$$x = \frac{1}{2}[(x+y) + (z+x) - (y+z)]$$

is also a rational number, which contradicts the fact that x is an irrational number. Next, prove that the two-color complete graph K_5 contains no blue pentagon. Otherwise, suppose that there is a blue pentagon $xyzuv$ which means that $x + y, y + z, z + u, u + v, v + x$ are rational numbers. Then

$$x = \frac{1}{2}[(x+y) + (z+u) + (v+x) - (y+z) - (u+v)]$$

is still a rational number, which contradicts the fact that x is an irrational number. The two-color complete graph K_5 contains neither a blue triangle nor a blue pentagon. By the vice versa after Fig. 8.3, there must exist a red triangle. Suppose that $\triangle xyz$ is a red triangle, then $x + y, y + z, z + x$ are all irrational numbers.

12 As Fig. 16(1) shows us, color the K_7 in two colors and denote a solid line by a red edge and a dotted line by a blue edge. There are 4 red triangles: $\triangle A_1A_4A_6$, $\triangle A_2A_4A_6$, $\triangle A_3A_4A_6$, $\triangle A_7A_4A_6$ and 4 blue triangles $\triangle A_1A_2A_3$, $\triangle A_2A_3A_7$, $\triangle A_1A_3A_7$, $\triangle A_1A_2A_7$. It is easy to see that any two monochromatic triangles with the same color contain a common edge. So $n \geqslant 8$.

Next, we prove when $n = 8$, the proposition is true. We prove by contradiction.

First we prove a lemma: If the proposition is not true, there must be a red triangle and a blue triangle which contain one common vertex. First, K_8 which is colored in 2 colors must contain a monochromatic triangle. Without loss of generality, let it be a blue triangle $\triangle A_1A_2A_3$. Now among $A_3A_4A_5A_6A_7A_8$, there must exist a monochromatic triangle which is not blue. If this red triangle contains A_3, the lemma is true. Otherwise, suppose that $\triangle A_4A_5A_6$ is a red triangle. There are 9 edges joining $\triangle A_1A_2A_3$ and $\triangle A_4A_5A_6$ among which there are at least five monochromatic edges. Without loss of generality, let it be red. Then A_1, A_2, A_3 are incident to at least 5 red edges. Among them there must be one vertex which is incident to at least two red edges. This triangle consisting of this vertex and $\triangle A_1A_2A_3$ contains one common vertex. So the lemma must be true. Next, we prove the proposition: Suppose the proposition is not true. By the lemma, let $\triangle A_1A_2A_3$ be a blue triangle and $\triangle A_3A_4A_5$ be a red triangle. Consider the edges joining $A_1A_4A_6A_7A_8$. Among them there is no monochromatic triangle. So K_5 consists of a blue cycle of 5 vertices and a red cycle of 5 vertices (by the remark after Fig. 8.3). In Fig. 16(2), represent red edges by solid lines and blue edges by dotted lines. Without loss of generality, let $A_1A_4A_6A_7A_8$ be a blue cycle of 5 vertices and $A_1A_7A_4A_8A_6$ be a red cycle of 5 vertices. Next, we discuss the color of A_3A_7. If A_3A_7 is blue, then A_3A_8, A_3A_6 must be red. (Otherwise, $\triangle A_3A_7A_8$ or $\triangle A_3A_6A_7$ is blue, which contradicts $\triangle A_1A_2A_3$.) Now $\triangle A_3A_6A_8$ is red, which contradicts $\triangle A_3A_4A_5$! If A_3A_7 is red, we discuss the color of A_3A_8.

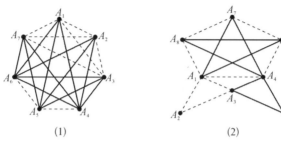

Fig. 16

If A_3A_8 is blue, A_2A_4 must be red, A_2A_8 must be blue and A_2A_7 must be red. So, $\triangle A_2A_4A_7$ is a blue triangle, which contradicts $\triangle A_3A_4A_5$. If A_3A_8 is red, A_5A_7 must be blue and A_5A_8 must be blue. So $\triangle A_5A_7A_8$ is blue, which contradicts $\triangle A_1A_2A_3$! In summary, when $n = 8$, the proposition is true. So the smallest natural number is 8.

13 We denote 20 teams by 20 vertices. Join the vertices of the teams which have played matches in the first round by red edges and the vertices of the teams which have played matches in the second round by blue edges. Then every vertex is incident to a red edge and a blue edges. The graph must consist of several even cycles. In every even cycle, we can choose half of the vertices among which any two vertices are not adjacent. Then we choose 10 teams which have not played with each other. We complete the proof.

Exercise 9

1 (1) When $n = 5$, Fig. 17 is what we need. When $n = 6$, Fig. 18 is what we need.

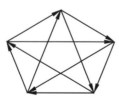

Fig. 17

Fig. 18

(2) Suppose when $n = k$, there exists a directed graph satisfying the requirement. When $n = k+2$, first using the vertices $V_1, V_2, \ldots, V_k$, we draw a directed graph with k vertices satisfying the requirement. For another two vertices V_{k+1}, V_{k+2}, suppose that $V_1, V_2, \ldots, V_k$ all point to V_{k+1} and that V_{k+2} points to $V_1, V_2, \ldots, V_k$. Suppose that V_{k+1} points to V_{k+2}. Then V_{k+1} gets to $V_1, V_2, \ldots$ through V_{k+2}. (Clearly, V_{k+2} can get to $V_1, V_2, \ldots, V_k$.) $V_1, V_2, \ldots, V_k$ can get to V_{k+2} through V_{k+1}. (Clearly, $V_1, V_2, \ldots, V_k$ can get to V_{k+1}.) So this graph with $k+2$ vertices still satisfy the requirement.

By (1) and (2), we know for any $4 < n \in \mathbf{N}$, there exists a scheme of changing the path among n cities satisfying the requirement.

2 Suppose G contains a circuit $(v_1, v_2, \ldots, v_k)$. In $v_2, v_3, \ldots, v_{k-1}$, take the first vertex v_i so that the arc (v_{i+1}, v_1) exists. Then there exists an arc (v_1, v_i), so (v_1, v_i, v_{i+1}) is a triangular circuit.

3 We will prove that if an air route satisfies the condition f and there is no flight between two cities A and B, then we can use airline $A \to B$ or $B \to A$ so that the air route still satisfies the condition. If not, the new route does not satisfy the condition f. Then after opening the route $A \to B$, there exists a closed path $B \to C_1 \to \cdots \to C_n \to A \to B$. Similarly, after opening the route $B \to A$, there exists a closed path $A \to D_1 \to \cdots \to D_m \to B \to A$. But before opening route between A and B, there exists a route $A \to D_1 \to \cdots \to D_m \to B \to C_1 \to \cdots \to C_n \to A$. (Maybe there are some vertices C_i and D_j which are overlapped. It means that the former air route does not satisfy the condition f, because it is possible to fly from A and return to A. A contradiction.)

4 Refer to Example 4.

5 We denote n players by n vertices. If v_i defeats v_j, we can draw an arc from v_i to v_j to get a directed graph D. If there is no circuit in D, there must exist a vertex v whose indegree is 0. The vertex represents the person who wins all the games. Similarly, we can prove there is a person who loses all the game.

6 Suppose among $v_1, v_2, \ldots, v_n$, the vertex v_p has the most number of offsprings, then v_p is the common ancestor of the n persons. Otherwise, we assume that v_p is not the ancestor of v_q, and then the common ancestor v_r of v_p and v_q is not v_p, and the offsprings of v_r are more than those of v_p by one. A contradiction.

7 B has won two games.

8 One round rohin corresponds to one tournament. By assumption, there is no vertex whose outdegree is $n-1$. By the Pigeonhole Principle, there exists at least two vertices whose outdegrees are the same. By Theorem 4, we have completed the proof.

9 Use Theorem 3.

Index

Adjacent, 3
Arc, 101

Bigraph, 24
Brouwer, 57

Chain, 40
Circuit, 103
Coloring, 84
Complete graph, 3
Completet k-partite graph, 24
Connected graph, 40
Cycle, 40

Degree, 13
Dirac, 70
Directed graph, 101
Direct search, 63
Endpoint, 101
Euler's Formula, 75
Euler tour, 53
Eulerian, 53
Even, 13
Extremal graph, 24

Face, 75

Finite graph, 3
Forest, 41

Graph, 1
Generating tree, 45

Hamiltonian cycle, 63
Hamiltonian graph, 63
Hamiltonian path, 63, 106
Hamiltonian, 63
Hand-Shaking lemma, 14
Homeomophic, 78
Hypergrah, 94

Incident, 3
Indegree, 102
Infinite graph, 3
Isomorphic graph, 2

Königsberg, 51

Leaf vertex, 41
Length, 40, 103
Loop, 3

Maximum degree, 13

Minimum degree, 13

Odd, 13
Ore, 68
Outdegree, 102

Parallel edges, 3
Path, 40
Pendant vertex, 41
Peterson graph, 132
Planar graph, 75

Ramsey number, 90

Regular graph, 13

Schur Theorem, 95
Simple graph, 3
Starting point, 101
Subgraph, 2

Tournament, 102
Tree, 41
Triangle, 25
Turán's Theorem, 28

Vertex, 1